TRUE WARNINGS AND FALSE ALARMS

Evaluating Fears about the Health Risks of Technology, 1948–1971

ALLAN MAZUR

RESOURCES FOR THE FUTURE
Washington, DC, USA

An RFF Press book
Published by Resources for the Future
1616 P Street NW
Washington, DC 20036–1400
USA
www.rffpress.org

Library of Congress Cataloging-in-Publication Data

True warnings and false alarms : evaluating fears about the health risks of technology, 1948–1971 / Allan Mazur.
p. cm.
Includes bibliographical references and index.
ISBN 1-891853-55-4 (hardcover : alk. paper) -- ISBN 1–891853–56–2 (pbk. : alk. paper)
1. Health risk assessment—History—20th century. I. Title.
RA427.3.M39 2003
610—dc21 2003013581

f e d c b a

The paper in this book meets the guidelines for permanence and durability of the Committee on Production Guidelines for Book Longevity of the Council on Library Resources.

This book was designed and typeset in Giovanni Book by Grammarians, Inc. The cover was designed by Rosenbohm Graphic Design. The text is printed on recycled paper.

The findings, interpretations, and conclusions offered in this publication are those of the author. They do not necessarily represent the views of Resources for the Future, its directors, or its officers, and they do not necessarily represent the views of any other organization that has participated in the development of this publication.

ISBN 1-891853-55-4 (cloth) ISBN 1-891853-56-2 (paperback)

About Resources for the Future *and* RFF Press

Resources for the Future (RFF) improves environmental and natural resource policymaking worldwide through independent social science research of the highest caliber. Founded in 1952, RFF pioneered the application of economics as a tool for developing more effective policy about the use and conservation of natural resources. Its scholars continue to employ social science methods to analyze critical issues concerning pollution control, energy policy, land and water use, hazardous waste, climate change, biodiversity, and the environmental challenges of developing countries.

RFF Press supports the mission of RFF by publishing book-length works that present a broad range of approaches to the study of natural resources and the environment. Its authors and editors include RFF staff, researchers from the larger academic and policy communities, and journalists. Audiences for publications by RFF Press include all of the participants in the policymaking process—scholars, the media, advocacy groups, NGOs, professionals in business and government, and the public.

CONTENTS

PREFACE

No one doubts that mistakes involving the use and application of technology were made in the United States during the twentieth century, but there is less agreement on what those mistakes were. News reports about technological error seem inevitably to be followed by rebuttals saying that the health risk was not a risk at all, or at least that it was not as bad as first reported. These conflicting accounts usually involve one expert's word against another's, without any definitive adjudication. When warned that genetically modified food, or smallpox vaccinations, or estrogen replacement for postmenopausal women is hazardous, how do we know whether or not to believe it? If we cannot properly assess the degree of a given risk, how can we learn from our mistakes?

I hope in part to remedy this ambiguity and to distinguish true warnings about technology from false alarms by analyzing 31 warnings—some true, some false—brought to public attention between 1948 and 1971. This odd range of years was dictated by Edward Lawless, who in 1977 published a collection of warnings from the news of that period, intended as a representative sample of the kinds of alarms then being raised about hazardous or otherwise problematic technologies. Lawless, trained as a chemist, was at the time of the study head of the Technology Assessment Section of the Midwest Research Institute's Physical Sciences Division. Lawless's sample does not exactly suit my purpose, as it includes some instances of technological fraud (not involving health risks), and it mixes allegations of environmental degradation with warnings about health effects. I limit my study to warnings about human health problems. Warnings about diffuse or poorly defined environmental effects, although important, add a level of difficulty that I avoid here. A few of Lawless's cases stretch common notions of "technology"; for example, he discusses a warning about botulism from contaminated cans of soup, but since he saw fit to include such cases in his sample I have retained them. My intention is, first, to explain why these warnings were difficult to evaluate when they were initially voiced and, second, to offer guidelines that may be helpful for evaluating future warnings. I make few explicit judgments about the truth or falsity of warnings currently in the news, although my general conclusions carry inferences of that kind.

My work on this book was supported by the National Science Foundation under Grant No. SBR-9808684 to Allan Mazur for the re-evaluation of public warnings raised during the 1950s and 1960s about technological and environmental hazards to health. Any opinions, findings, and conclusions or recommendations expressed here are my own and do not necessarily reflect the views of the National Science Foundation or of any other person or institution named here.

During the preparation of this book, I enjoyed the excellent hospitality of Resources for the Future in Washington, DC, which supported me with its Gilbert White Fellowship for 2000-2001. I am continually grateful for my permanent home at the Center for Environmental Policy and Administration in the Maxwell School of Syracuse University and to CEPA's director, W. Henry Lambright. Of the many people who helped me during this project, I especially thank Jennifer Bretsch, Richard Brickwedde, Lee Clarke, Terry Davies, David Driesen, Thomas Field, Edward Groth III, Kevin Jacobson, Raymond Letterman, Brian Martin, Rachel Mazur, Ernest Newbrun, Don Reisman, Kristin Shrader-Frechette, Jeffrey Stine, Albert Teich, and James Wilson.

Sections of Chapters 3 and 4 are adapted from two of my earlier books, *A Hazardous Inquiry* (1998) and *Dynamics of Technical Controversy* (1981). Appendix 2 comprises four articles previously published in the journal *Risk: Health, Safety & Environment.*

CHAPTER 1

TRUE WARNINGS AND FALSE ALARMS

Cries of alarm are raised so often over new technological risks that journalists, politicians, and the public have thrown up their collective hands in frustration over their inability to distinguish real hazards from false alarms. Which warnings deserve costly governmental solutions? Which should be provisionally ignored?

Public policy decisions in industrial societies increasingly focus on possible technological and environmental threats to human health (Beck 1992; Stern and Fineberg 1996). These may accompany the introduction of new technologies, like the fear of harmful radiation from proliferating cell phone towers. They may be late-recognized consequences of old technologies, such as the increased incidence of breast cancer now associated with estrogen replacement therapy. Engineering "solutions" thought to alleviate one hazard may introduce another, as when the chlorination of drinking water, which effectively stemmed gastrointestinal illness, was discovered to produce carcinogenic byproducts (Graham and Wiener 1995). With technology continuing to advance and intrude into more diverse domains, risk issues become ever more salient.

Often, if the underlying science has enough time to mature, we can reach confident conclusions about alleged risks. Today we know that chlorofluorocarbons really are damaging the stratospheric ozone layer. Sixty-hertz electromagnetic fields from transmission lines probably are not causing significant increases in leukemia. Human actions almost certainly have raised global temperatures. Silicone breast implants do not cause degenerative disease.

But new risks appear faster than we can scientifically evaluate the old ones. Are endocrine disrupters, the class of synthetic chemicals that possibly distort normal hormonal processes in our bodies, a cause for concern? What about genetically modified foods? Do cell phones cause brain cancer? Does their use while driving increase the likelihood of traffic accidents? Will resistant strains of malaria soon overwhelm the drugs available to attack the microbe? Are we changing people's personalities in harmful ways by overprescribing Prozac for adults and Ritalin for children?

Here, I address a specific problem in risk policy: How can we determine, at a fairly early stage, whether or not to take a new warning seriously? Early recognition is important because of the potential risk to our health and wealth. If we act quickly and the warning turns out to be false, money is wasted, people are needlessly frightened, and risk regulators lose their credibility. If we act too slowly and the hazard is real, we suffer environmental damage or detrimental health effects, perhaps even catastrophic losses.

This policy dilemma leads to an empirical question: Are there hallmarks of early warnings that predict when alleged risks will ultimately turn out to be true hazards rather than false alarms? The identification of such hallmarks is my goal here. Such benchmarks may not allow us to make perfect predictions about which warnings we should heed and which we should ignore. However, the hallmarks may reduce the number of times we respond late to true hazards, and reduce the time and money we devote to false alarms.

Fortunately, Edward Lawless (1977) compiled an excellent collection of public warnings from 1948–1971 for a study of products and processes that some people thought posed serious hazards to the public or the environment. Today, 30–50 years later, we can evaluate whether these warnings identified serious risks or were false alarms.

Lawless never intended his compilation for the purpose to which I put it here, but through his collection he has become my steady companion throughout this project. Since his work appears recurrently in these pages, I introduce him at the outset.

THE LAWLESS COLLECTION

In 1977 Edward Lawless published *Technology and Social Shock*. The book presents numerous contemporary case studies of public concern about technological and environmental hazards. It describes "episodes of public alarm over technology [and the environment]—'social shocks'—that have inspired major news stories in the media in recent years" (xi). The book was "viewed as a pioneering effort, a forerunner of many further studies and more intensive analyses" (xi).

Lawless and his team developed more than 40 detailed case studies, intending them as a representative sample of the kinds of technical concerns then in the news. Some are dramatic, some mundane. Three involve reproduction and genetics, 12 are about food and medicines, 16 involve radiation and environmental problems, and several concern projects of the federal government, particularly related to military technology. A few cases are brief historical notes about past technological and medical problems (Krebiozen, the supposed cancer cure) or environmental problems without serious threats to human health (corn leaf blight), or purely intellectual controversies (the authenticity of an antique statue). Dropping these cases reduces the number of relevant health warnings to 31 (Table 1-1).

A typical case is 10 pages long and follows a uniform format. Each case includes historical background, the key events that raised the alarm, the major actors, the disposition (if any) in terms of public or private policy, a summary evaluation, and a bibliography. These 31 case studies of alleged hazards are the basis for this book (See Appendix 1 for a summary of them.)

Table 1-1. Selected Substances and Processes Generating Public Health Warnings, 1947–71

Oral contraceptives	DMSO (dimethyl sulfoxide)	Enzyme detergents
Contaminated cranberries	Shoe fluoroscopes	NTA (nitrilotriacetic acid) in detergents
DES (diethylstilbestrol) in livestock	Medical x rays	Plutonium at Rocky Flats, Colorado
Cyclamate	Radiation from defective televisions	Radioactive waste stored in Kansas
MSG (monosodium glutamate)	Smog in Donora, Pennsylvania	Nuclear test on Amchitka Island, Alaska
Botulism	Mercury pollution from industry	Poison gas released at Dugway Proving Ground in Utah
Fish protein concentrate	Mercury in tuna	Nerve gas disposal
Fluoridation of water supply	DDT (dichloro diphenyl trichloroethane)	ELF (extremely low-frequency radiation) at Project Sanguine
Salk polio vaccine	Asbestos	Chemical mace
Thalidomide	Taconite pollution	Injuries on synthetic turf
HCP (hexachlorophene)		

Source: Lawless 1977.

RETROSPECTIVE RISK ASSESSMENT

I ask if, from today's perspective, each warning proved true or false. (This is not the same as asking if the warning was wise or foolish at the time, a question to which I return in the last chapter.) I approach this question as a sociologist and a technical generalist, making no attempt to substitute my own assessment for that of experts in the field of the alleged hazard. I obtained these expert assessments from the scientific and government literature and from interviews.

No scientific judgment is absolute. Objections or qualifications can be raised for every risk assessment (see Chapter 4). The technical controversy over fluoridation, for example, has continued for decades. Some competent scientists still insist that adding fluoride to drinking water is imprudent and without beneficial effect, while most experts maintain the practice is not harmful and reduces the incidence of dental cavities. If there is still considerable scientific uncertainty or dispute over the validity of a warning, I consider appraisals from opposing experts. However, I have not taken the existence of dissenting opinion per se to mean that the validity of an alleged hazard has not been determined. With sufficient searching, one can find experts to defend virtually every one of the Lawless warnings, whether because they really believe it or because they want to show intellectual virtuosity. I avoid agnosticism on that basis alone. In general, I have accepted recent scientific positions of orthodox and authoritative bodies such as the National Academy of Sciences, even when there are some dissenters.

For some cases, there remains sufficient scientific uncertainty that warnings still cannot be judged true or false on empirical grounds. For example, a half-century after the first warnings about very low levels of ionizing radiation, the harmful effect, if any, of low doses of x rays remains unknown (see Chapter 4). Nonetheless, it has become accepted regulatory wisdom—virtually without dissent—that human exposure to x rays should be avoided unless there are overriding benefits. This is the basis for the U.S. Nuclear Regulatory Commission's requirement that public exposure to sources of ionizing radiation shall be "as low as reasonably achievable" (Title 10, Code of Federal Regulations, Part 20). In the absence of empirical evidence, I take such a regulatory posture as a societal judgment that a warning is or is not valid from today's perspective. Thus, even without compelling empirical evidence, I regard early warnings against unnecessary exposure to x rays as valid. (See Chapter 6 for a full description of the basis on which I coded the warnings as true or false.)

HYPOTHETICAL HALLMARKS

My search for hallmarks of true-versus-false warnings is guided by four hypotheses. Each suggests a potential clue to the validity of an early alarm.

- True warnings are more likely than false alarms to reach the news media from reputable scientific sources operating in conventional scientific ways.

- False alarms are more likely than true warnings to have sponsors with biases against the producer of the alleged hazard.
- Hyped media coverage is more likely to indicate a false alarm than a true warning.
- A "derivative warning"—one arising from a popular social issue—is more likely to indicate a false alarm than a true warning.

True Warnings Are More Likely than False Alarms to Reach the News Media from Reputable Scientific Sources Operating in Conventional Scientific Ways

My first hypothesis speaks to the source of the first public warning—the person or organization providing information to the first journalist reporting about an alleged hazard. This hypothesis represents an "internalist" view of science; that is, the view that despite personal failings and biases among scientists, the conventional scientific enterprise usually produces reasonably objective information about hazards. According to this view, science eventually resolves many technical controversies through the weight of evidence and sound reasoning (Engelhardt and Caplan 1987). While unorthodox views of maverick scientists occasionally turn out to be correct, even revolutionary, more often they are wrong.

Early warnings about fluoroscopic shoe-fitting machines—x ray machines used to measure people's feet before they purchased shoes—were based on widely accepted scientific evidence about the cancer-causing properties of high doses of ionizing radiation. In contrast, early warnings about fluoridation came from politically active laypeople in Wisconsin objecting to "rat poison" in the water (Mazur 1981).

Rather than attempt to judge on substantive grounds whether or not an early (often contentious) warning was scientifically sound at the time it was made, I adopt the internalist perspective that valid science is conducted by professional Ph.D. or M.D. level scientists working in recognized research institutions and reported in recognized publications, usually after peer review. Thus I code public warnings according to whether or not they are based on reputable science.

False Alarms Are More Likely Than True Warnings to Have Sponsors with Biases against the Producer of the Alleged Hazard

This and the remaining hypotheses reflect "externalist" thinking; that is, technical claims are strongly influenced by the biases and self-interests of their proponents (Martin 1991; Krimsky and Golding 1992). People who dislike a corporation are likely to perceive its products and processes negatively, even as hazardous, whereas company executives see them in a positive light. Warnings originating with disinterested parties are, according to this hypothesis, more likely to be valid than those originating with individuals or groups with a history of antipathy toward agents promoting or profiting from the alleged hazard. By the same token, assurances coming from the producer of a suspected hazard are less credible than those from an independent expert.

Personal biases may remain invisible, but we can often identify in professional affiliations and biographical material clear indications that a warning is correlated with a broader pattern of hostility against the source of the alleged hazard. For example, prominent opponents of fluoridation carried strong ideological biases against big government and big corporations. They were especially biased against the aluminum industry, which they regarded as foisting fluoride compounds, waste products from aluminum manufacture, onto unsuspecting communities (Mazur 1981). In contrast, warnings about fluoroscopic shoe-fitting machines came from scientists with no prior grievances against manufacturers of shoes or fluoroscopes.

Hyped Media Coverage Is More Likely to Indicate a False Alarm Than a True Warning

In many cases, public and governmental concerns over an alleged hazard increase or decrease with the amount of media attention. Unfortunately, the amount of news coverage given to various hazards bears little relation to their measurable or calculable risks (Mazur 1984, 1987; Singer and Endreny 1994). Perhaps, then, hints at the validity of an early warning may be found in the manner with which it is covered in the news.

Media coverage of each warning is coded as routine or hyped (see Appendix I). Routine coverage occurs when the story receives an ordinary amount of space and salience, and continues for a common amount of time. Hyped coverage occurs when a source or journalist uses unusual means—including sensational or clearly exaggerated information or extraordinary visuals—to heighten news coverage or intensify alarm. For example, the warning against cyclamate, an artificial sweetener, is coded as hyped because it became a public issue when a mid-level scientist from the Food and Drug Administration appeared on NBC television news displaying deformed chicken embryos produced when cyclamate was injected into eggs. This was followed by a special press conference called by the U.S. Secretary of Health, Education, and Welfare. In contrast, the warning against shoe fluoroscopy became public through ordinary news reports based on two articles in the *New England Journal of Medicine* in 1949; it is coded routine.

Obviously, there is a subjective element in judging whether news coverage is sensational or not. Fortunately, Lawless provides some helpful guidelines. In many cases he notes that journalistic attention was extraordinary. He also provides a quantitative measure of media coverage for each warning. While Lawless's own judgments are not free from bias, at least he made them without knowledge of my hypotheses. Just as important, Lawless noted the character of media coverage without the 30 years of evidence I used to determine whether the warnings proved valid.

A "Derivative Warning"—One Arising from a Popular Social Issue—Is More Likely to Indicate a False Alarm Than a True Warning

Certain types of hazards are topical one year, out of fashion the next. Examples include chemical waste, global warming, nuclear power plant accidents, and radon. These issues held no interest for the news media during certain periods, then became of intense interest, then fell from favor again (Mazur 1984, 1987, 1998; Mazur and Lee 1993).

Some of these warnings seem to rise in journalistic attention through their connection to other (often bigger) news stories, like a surfer riding a wave. Warnings against fluoridation, with its purported implications of socialized medicine, stemmed largely from broader concerns about socialism and communism in the United States. Fluoridation alarms rose and fell with national interest in this larger political concern (Mazur 1981), suggesting that risk per se was not the essential driving force behind the alarm. Perhaps warnings that reach the news on their own merits are more likely to be valid than warnings covered because of their connection to collateral issues of interest.

STRUCTURE OF THE BOOK

These four hypotheses guide my analysis of Lawless's cases. To the extent that any of them are verified, they offer early hints that a warning is true or false.

Because nearly all the warnings that Lawless examined were scientifically contentious in their day (some still are), it is worth trying to understand why experts disagree over basic scientific facts; this is the subject of Chapter 4. Chapter 5 explains the procedures I used to retroactively assess whether the warnings Lawless examined were true or false. In Chapter 6 I explain how I further classified Lawless's warnings so I could test my hypotheses concerning the hallmarks of true and false warnings. Chapter 7 contains my evaluations of these hypotheses, identifying specific characteristics that often differentiate true warnings from false alarms. In Chapter 8 I explore changes in the risk landscape since Lawless's era to shed light on whether the hallmarks I have discovered apply to more recent technological warnings.

Although my analysis treats Lawless's warnings as 31 separate cases, they must be understood as elements of our turbulent postwar culture. Therefore in Chapter 3, I embed all 31 cases in a brief history of the period 1948 to 1971, connecting them to other important technological and environmental issues of the time. But first, in Chapter 2, I sketch a history of American views of technology from the beginning of the century through World War II, setting the stage for Lawless's era.

CHAPTER 2

TECHNOLOGY AS FRIEND AND SOMETIME FOE: 1900–1947

Technology is causing unprecedented social change, claimed Alvin Toffler in his 1970 bestseller *Future Shock*. Our failure to adjust, he thought, produced widespread discontent. Toffler notwithstanding, the undeniably rapid changes during the 1960s were nothing compared with the social and technical transformations that had occurred around 1900.

Overall, Americans expressed few worries about the technologies proliferating at the beginning of the twentieth century, even though these innovations were profoundly changing their lives and their environment. Newspapers and magazines expressed considerable admiration for these developments and carried copious advertisements for new products, raising few serious warnings. Industrial disputes focused on wages, working conditions, and profiteering, but people evinced little worry about the products or processes themselves.

The sharpest opposition to technology had arisen in the nineteenth century among Plains Indians, who objected that railroads built on their lands had destroyed their subsistence. Buffalo herds would not cross the tracks, which disrupted their migration patterns. Riflemen traveling by train hunted the animals nearly to extinction. But by 1900 the Indians and their protest had been suppressed, leaving few declared opponents of technology.

To appreciate the turnaround in American attitudes toward technology after World War II, one must understand the reasons for American enthusiasm during the first decades of the century, as well as the strains and ambivalence reflected in the turn-of-the-century's conservation movement and during the Great Depression. This chapter provides a historical prologue to Lawless's era of technological warnings.

MARVELS OF 1900

The early twentieth century saw the diffusion of telephones, automobiles, tractors, electricity, indoor plumbing, aspirin, radio, and movies. American cities were transformed by early skyscrapers. Millions of immigrants arrived cheaply and swiftly on steamships, which had recently replaced sail vessels. Blimps and airplanes, though not available for use by most people, were visible in the sky. Out of view at home, though still a source of national pride, was the greatest technological feat: the Panama Canal. Cut by American engineers, after failure by the French, the canal opened for traffic in 1914. It was one more triumph in a time of optimism and seemingly endless potential for progress (Hughes 1975).

Of course, since the beginnings of the Industrial Revolution, critics had proffered a long list of complaints: factory pollution, the elimination of jobs by machines, the monotony of the assembly line, the rise of materialism, the blandness of standardization, the conflict between science and religion. These concerns, always greater in Europe than America, did not negate the more common impression on both sides of the Atlantic that technology was wonderful, its future unlimited.

World War I challenged these views, though again far more in Europe than in the United States. Mechanized killing wasted a generation of Europeans, while Americans suffered far less and remained nearly as optimistic as ever. With the upswing of the 1920s, even Bertrand Russell, the

British philosopher and critic, put forward an essentially positive view: "I have not sought to minimize what may be considered the defects of the machine civilization. I do not doubt, however, that its merits far outweigh its defects. Take two items alone: the diminution of poverty and the improvement in public health. These two alone represent an almost incalculable increase in average happiness" (1928, 81).

Signs of industrialization and optimism can be seen in Sears mail order catalogs. The cover of the 1900 catalog shows agrarian America, with a farm and a church under a blue sky and factory chimneys barely noticeable in the background. Industry is front and center on the 1923 cover, which features an ocean liner puffing brown smoke, a truck, a train, a crane, and skyscrapers. Inside the 1900 catalog are kerosene lamps, gramophones, coal stoves, horse buggies, and magic lanterns (a type of early slide projector, used as entertainment before movies became popular). The 1923 catalog offers sewing machines, cameras, gas stoves, central heating, indoor plumbing, electric lighting, automobile parts, telephones, gasoline engines, and appliances such as vacuum cleaners, irons, toasters, and even vibrators. Aimed not at rich people but at ordinary Americans, the Sears catalogs depicted major transformations in daily life.

THE CONSERVATION MOVEMENT

Many of the warnings studied by Lawless reflected the forceful growth of American environmentalism during the 1960s. This "new" movement was firmly rooted in the conservation movement of the late nineteenth and early twentieth centuries, which lay dormant between the world wars. In the decades after World War II, old-line conservationist organizations such as the Sierra Club and the Audubon Society blossomed again and were joined by a spate of new organizations.

While modern environmentalism is sometimes characterized as antitechnology, precursor conservationists are rarely described that way. Some of the leading personalities, most famously Theodore Roosevelt, were devotees of modern technology. But certainly the early movement was a reaction to rampant industrialization: amid their marvels, modernizing cities and mushrooming factories had distinctly unpleasant aspects.

Crowded slums spawned crime and disease; skies and waterways became polluted. Many people with money moved to quiet suburbs with lawns, but the problems of urban life were unavoidable unless one traveled even farther away. People of wealth and education had the means and increasingly the inclination to visit the country, even the wilderness, as tourists, hunters, and fishers, recharging themselves to cope with urban life.

The wilderness had rarely appealed to people in agrarian societies, who associated wild places with real or imagined hazards to themselves and their crops and livestock. This view was changing by the nineteenth century, when it became fashionable for privileged Europeans to visit the Alps or Britain's Lake District, places cherished by the Romantic poets and artists for their untamed scenery. American landscape painters of the Hudson River School elaborated this theme, showing the sublime beauty of the Catskill Mountains as if endowing the nation's virgin forests, mountains, and streams with moral and religious virtue.

Americans went further than Europeans, not only idealizing wilderness areas but also taking strong measures to preserve them. Protecting nature from industrialization and development is an indigenous impulse, the only response against technological encroachment that was stronger in the United States than in Europe before 1950. Europeans had been cultivating their land for centuries; little was undeveloped by 1900. The United States still had a vast wilderness with unsurpassed scenery. After the Civil War, much of this land became accessible via the growing network of railroads.

The grandeur and degradation of the American landscape and attempts to protect it are illustrated by Niagara Falls in the East and the Sierra Nevada in the West. Frederick Church's 1856 rendering of Horseshoe Falls, perhaps the most popular American landscape painting of the nineteenth century, helped establish Niagara Falls as a prime tourist destination. But the waterfall's sublimity was punctuated by the occasional daredevil walking across it on a tightrope or plunging down in a barrel, and Canadian and U.S. shorelines, held in private hands, became centers of commercialism and hucksterism, disfigured by mills, hotels, honkytonks, and billboards. A preservation campaign led by landscape architect Frederick Law Olmstead convinced New York State in 1885 to purchase and clean up some of the U.S. shore—the first time a state had expropriated land for purely aesthetic reasons. Canada took similar action two years later. But these cures for the shoreline were only partial, and new threats were emerging.

Electrical generation and electrochemical companies were attracted to Niagara like iron filings to a magnet. Promoters saw the energy of copiously falling water, wasted on rocks below, as a potential source of profit or societal improvement. King Gillette, inventor of the safety razor, proposed the most grandiose scheme, a utopian city of 60 million inhabitants, its huge skyscrapers and factories covering the gorge and electrified by the falls. Placing nearly the entire U.S. population in one location, Gillette thought, would allow better coordination of work and more efficient distribution of goods and services.

By 1900 upstream water was being routed out of the river into downward sloping tunnels, where it turned electrical turbines before returning to the river below the falls. Canadian and U.S. interests each tried to maximize their share of the rushing water. Technological enthusiasts like writer H.G. Wells and physicist Lord Kelvin thought it acceptable to divert water from the precipice if its power could be used for social betterment (Berton 1992).

A new campaign by preservationists in 1905 was less about blighted shores than about saving the waterway, which was in real danger of becoming a trickle. Again there was partial success, with an international agreement limiting the amount of water each side could draw. Yet withdrawals continued over the years, at times amounting to half the total flow.

The area saw little scenic improvement until after World War II, when additional shoreline was converted to park, though more successfully on the Canadian side. Visitors today are still impressed by the grandeur of Niagara Falls, but the commercial surroundings are inescapable, and blighted areas remain.

Niagara Falls was far more accessible to most Americans than California's Sierra Nevada, which white settlers had barely penetrated before gold was discovered there in 1848. The "forty-niners"—arriving a year later—found spectacular valleys and enormous trees. Bark stripped from one sequoia was sent for exhibit back East, where skeptics thought it too large to be genuine. News of the trees and pictures of Yosemite Valley raised the tourism value of the region while mining operations were ruining the landscape. Rivers were dammed and diverted for miles

The General Noble Tree was felled in 1892 for display at the World's Columbian Exposition in Chicago. The trunk was cut 50 feet above its base, hollowed out, disassembled into sections for transport, and then reassembled at the fair. Photo by C.C. Curtis, courtesy of the National Park Service, Archives of Sequoia and Kings Canyon National Parks.

so miners could sift for gold in the sand of their beds. Networks of flumes and pipes collected water from the melting snow pack, directing it under high pressure against entire hillsides, washing away tons of dirt to reach grains of gold. In addition to damaging the scenery, hydraulic mining dumped millions of cubic yards of debris into the rivers of the Sacramento Valley, causing flooding and silting and destroying farms. Damage was so great that hydraulic mining was outlawed in 1884 (Webster 1972), by which time protectionist forces were making some gains. In 1872 Congress created the world's first national park, Yellowstone; in 1890 three more followed in the Sierra Nevada: Yosemite, Sequoia, and General Grant (the forerunner of Kings Canyon).

In 1903 Yosemite was the site of a famous encampment where naturalist John Muir and President Theodore Roosevelt slept in the snow, savoring the magnificent setting. Muir had founded the Sierra Club (1892), dedicated to preserving the Sierra Nevada wilderness. He had tramped the mountains for years, describing his travels in his eloquent writings as quasi-religious experiences, urging that such places of special beauty be protected from human touch. Roosevelt was an avid conservationist yet would build the Panama Canal, cutting a ditch though jungle to connect the Atlantic and Pacific oceans. He was a "man's man" who preached the virtues of a rugged outdoor life and loved to shoot large animals.

Less reverential than Muir about preservation, Roosevelt preferred the pragmatic view of conservation espoused by forester and conservationist Gifford Pinchot, who believed that natural resources should be protected but also made useful for a growing nation. Pinchot preached that forests could be harvested for wood and remain beautiful places for recreation if scientifically managed. For Pinchot technology was a friend if used wisely in husbanding natural resources but an enemy when used destructively. His theme of wise management was the intended basis for the National Forest Service, created by Roosevelt in 1905 with Pinchot as its first director.

The pure preservationist position of Muir came to be differentiated from the utilitarian conservationism of Pinchot, who sought a compromise between protecting the land and harvesting its bounty. The two men, initially allied against rampant destruction of the landscape, eventually fell out over the proposal to dam the Hetch Hetchy Valley, an area within Yosemite National Park and nearly as spectacular as Yosemite Valley. Muir and the Sierra Club strongly opposed the plan. Pinchot thought the project had sufficient merit to justify flooding the valley, sacrificing its scenery to progress. Hetch Hetchy was converted in 1913 to a reservoir for San Francisco and remains under water today.

Still, there was room in this vast country to set aside a few places of special scenic value. Although Muir lost Hetch Hetchy, his preaching of hands-off preservation rallied enough support to bring about the creation of the National Park Service in 1916, two years after his death, to protect specially designated areas and their wildlife. By that time, there were 14 national parks and 21 national monuments. One admirer, writer Wallace Stegner, called the new agency "the best idea America ever had." Perhaps it was. At least it was the acme of America's early attempts at environmental protection. With Europe engaged in the Great War, other matters would soon take priority.

Conservation did not re-emerge as a major public issue until 50 years later. When it did, in the 1960s, traditional issues were accompanied by new concerns, particularly radioactive fallout from nuclear weapons tests and the proliferation of chemical pesticides. These explicit warnings about specific technologies set the tone for many of the protests of Lawless's era.

TECHNOLOGY AS IDEOLOGY

The conservation ethic abated in the 1920s while technology forged ahead, even becoming an aesthetic ideal. American critic Lewis Mumford observed that "since, roughly, 1880, there has gone on a revival in typography, textiles, furniture, in architecture and city planning which shows, I think, that science and technics, while they have altered the basis of these arts, have not done away with the possibilities of their proper growth and development" (1928, 306).

Chicago's two World's Fairs, held in 1893 and 1934, dramatically illustrate the nation's changing aesthetics. Visitors to the 1893 Columbian Exposition delighted in the neoclassical architecture of its White City, with its stately buildings surrounding a placid lagoon. Though using modern construction methods and skeletons of iron and steel, these buildings, nearly all white, had facades in the styles of ancient Greece or Rome or the Italian Renaissance. In contrast, the 1934 Century of Progress was the first international exhibition dedicated to science. Its vividly colored pavilions were designed in the latest Art Deco style, reflecting the speed and efficiency of the Machine Age. One building was designed to look like a radio cabinet. Visitors could view the fair from rocket cars, made of aluminum and glass, suspended high overhead. Bright colored lights illuminated the fair at night, and searchlight beams of many colors reached across the sky (Allwood 1977).

During the first decades of the century, the combination of fast, safe elevators and structural steel allowed a continual increase in the height of new office buildings. After 1920 American architecture and design were influenced by Europe but in a modernist vein. The German Bauhaus movement was committed to the merger of arts and crafts with machine technology. Le Corbusier, the great Swiss architect, idealized the city as an efficient machine—shades of King Gillette. Technical themes were applied to the interiors as well as the exteriors of buildings and to furniture and decoration.

Growing, industrializing cities offered jobs and excitement. The jingle "How ya gonna keep em down on the farm, after they've seen Paree?" applied as well to New York, even to Buffalo. It was plain to nearly everyone that technology had fostered all this. Inevitably, engineers were portrayed as heroes of the age, as in this 1929 advertisement:

> When the clock hands meet at midnight he is still at work ... dreaming over streets and structures he will never live to see. He toils behind the scenes of great civil enterprises, the unsung prophet of comforts and economics which will bless the lives of generations as yet unborn. Yet the engineer must contend with the fantasies of idealists, the rhetoric of demagogues, the lobbying of propagandists. He must check every contingency of the future against the facts and figures of today. He must bring the cool wisdom of science to every choice of methods and materials (quoted in Petroski 1997, 63).

As engineers reshaped and controlled the natural environment, it did not seem much of a reach to control the social environment as well. Industrial engineer Frederick Taylor used time and motion studies to fit workers into the assembly line like cogs in a machine, suggesting that engineers had expertise in managing humans as well as mechanical elements of the system.

One stream of progressive thought emphasized centralized planning and efficient administration by experts. Economist Thorstein Veblen (1921) contrasted the rationality he saw in the mechanical industry with the waste and efficiency of business and finance. He concluded that engineers, with their superior technical skills, should take over management functions from business owners. Many commentators wrote of the aptitude of technical personnel to plan and operate an increasingly technical industrial society (Little 1924; Millikan 1930). The theme had become familiar by 1929, when Herbert Hoover, an engineer, was elected President.

THE GREAT DEPRESSION AND AMBIVALENCE

The Great Depression made technology problematic for Americans. Charles Beard wrote at the time, "The battle over the meaning and course of machine civilization grows apace, with resounding blows along the whole front .... A subject mildly discussed in women's clubs has broken into offices, factories, smoking compartments and political assemblies" (1930, 1). Many believed a massive failure of machine industry had caused the 1929 crash of the stock market. Others, among them L. Wallace, a prominent businessman, remained supremely confident about technology: "[I]n connection with the recent stock market debacle, [the President] held a series of conferences with the leaders of business and industry to determine what might be done .... When the facts were collected and analyzed, [he assured] the people of the nation that they might proceed about their daily tasks with confidence. [Would this] have been accomplished so effectively ... had the President been other than an engineer? (1930, 190).

The technocracy movement that bloomed in the early 1930s was symptomatic of this split view of technology as both curse and cure. According to the technocrats, increasingly efficient machines and assembly line techniques, which had reduced the need for human labor, were the prime cause of Depression-era unemployment. Influenced by Veblen, they also blamed the inefficiency of bankers and businessmen who manipulated the system for their own financial gains. The implied solution was government by apolitical technologists who would run society according to principles of scientific rationality and mechanical harmony (Arkright 1933; Elsner 1967; Akin 1977).

For all its enthusiasm, the rhetoric of technocracy was often incoherent. More articulate statements concerning the technological condition soon appeared, including the criticisms of Lewis Mumford in *Technics and Civilization* (1934) and Charlie Chaplin in *Modern Times* (1936). While these negative views found a receptive public, modernism flourished in architecture and product design, partly because some of the best European practitioners (including Walter Gropius and Mies van der Rohe) immigrated to the United States, but also because of increasing acceptance of technology as an aesthetic ideal.

WORLD WAR II AND THE REBIRTH OF CONFIDENCE IN TECHNOLOGY

Ambivalence about technology ended with the outbreak of World War II. Any technology was good if it aided the war effort: radar, code breaking, proximity fuses, submarine detection, jet engines, bomb sights, calculating machines, atomic fission. These innovations helped bring victory, a lesson that American and Soviet troops, closing in on Germany in the final days, had already learned as they raced to capture the best Nazi engineers for home use.

Morality in 1945 was unambiguous. It did not seem excessive to destroy two Japanese cities with one bomb apiece. The war had been long and bitter. In the United States racist notions about Asians were reinforced by atrocities committed by the Japanese on American prisoners. The tenacious resistance of Japan's soldiers during battles for the Pacific islands and the suicide attacks of kamikaze pilots diving explosive-laden airplanes into American ships fostered a belief that the Japanese would defend their homeland at any cost.

All sides targeted civilian populations. The Allies' fire-bombings of Dresden, Hamburg, and Tokyo had already killed tens of thousands of civilians. If the atomic bombs did not end the war, planners believed an invasion of Japan would be necessary. Thus, with the exception of some scientists who thought a nonlethal demonstration of an atomic explosion would convince the Japanese to stop fighting, most Americans found it acceptable to use the bombs on people.

The losers are blamed for every war. In this case, Germany bore the brunt. Today we know enough of Polish and Austrian complicity in the Holocaust—and we have seen enough new genocides—to understand that Germans of the Hitler era were not uniquely inhumane. Nonetheless, the Nazi extermination factories marked a low point in technological depravity that still eludes full comprehension. The United States, in sharp contrast, was at its finest hour. It had won an unqualified victory, suffered the fewest casualties among major combatants, and was unscathed on its home soil. Now far richer, more scientifically and technologically advanced, and more powerful than any competitor, than any nation *ever*, it had nearly reached the pot of gold at the end of the rainbow.

How quickly the rainbow faded. Old problems submerged by the war re-emerged, joined by new problems stemming from demobilization and America's leading role in world affairs. Technology seemed always to play a hand, usually as part of the solution but increasingly as part of the problem. The atomic marvels and other technologies that were admired for winning the war and were thought to have so much promise in keeping the peace themselves became the source of worry and insecurity. At first slowly, then more rapidly, public admiration for machines and chemicals and for engineers and scientists, which had barely been tarnished during the first half-century, would crumble during the following decades.

CHAPTER 3

LAWLESS'S ERA: 1948–1971

A rather new kind of political conflict can be seen emerging in newspapers after World War II: the technical controversy. At its core, such a controversy involved an esoteric though public dispute among experts about the extent to which an alleged hazard actually threatened human health or the environment. The warning that cigarettes cause lung cancer is an exemplar, breaking into the news in 1950 and remaining contested by tobacco interests until 1999. In the technical controversy, scientific findings and counterfindings, assertions and refutations are reported by the mass media to an often confused but concerned public. Legislators and jurists, faced with mediating and addressing such disputes, understand the problems hardly better than the public at large. Technical controversy, with all its opacity, had become a normal part of modern public discourse. It would pervade the warnings studied by Lawless, which I overview in this chapter.

Looking back at the technological optimism of the prewar decades, one finds occasional precedents for these postwar warnings about the effects of products and processes. During the 1920s, for example, scientists debated the wisdom of adding tetraethyl lead to gasoline (Warren 2000). More famous among historians of technology is the "battle of the currents" between George Westinghouse and Thomas Edison, over alternating current (AC) versus direct current (DC) as the best mode for electricity generation.

Edison was already selling DC electricity in 1888, when Westinghouse bought exclusive rights to the Tesla alternating current system. DC worked fine within a mile of a generator, but farther away light bulbs dimmed. In contrast, AC could be sent hundreds of miles with little loss of power. Edison insisted that AC was destructive to the nervous system and too dangerous to bring into homes. To illustrate, he made movies of animals killed in his laboratory with a thousand volts of AC (today we know that the lethal effect depends on high current, not high voltage). Westinghouse retorted that Edison was a liar and did not understand AC. At that time the New York State legislature was looking for a more humane way than hanging to execute criminals. Edison advised execution with 1,000 volts from an AC generator manufactured by Westinghouse. Edison's attorney suggested that the new electric chair be called "the Westinghouse." The first execution was badly botched, as the victim had to be electrocuted twice, but the innovation survived (although the attorney's suggestion did not). And AC, of course, became the accepted mode of generation (Adair 1996).

After World War II, technical controversies became more frequent and received far more attention from the news media, which were far more pervasive than in Edison's time. With the exception of intentionally polemical investigative reporters, journalists usually report controversies in a balanced manner, leaving readers and viewers to draw their own conclusions. Reporters give equal weight to both sides, even when one represents the scientific mainstream and the other an unusual view from one or a few dissidents. As news coverage of controversial issue increases, people become increasingly worried.

These worries were part of the malaise Alvin Toffler tapped when he published *Future Shock* in 1970. Within the receptive audience was Edward Lawless, who soon began to examine public warnings about technology, calling his cases "social shocks." All the warnings he studied first appeared in a major U.S. news medium between 1948 and 1971. Plotting new warnings over time shows a clear increase across the period (Figure 3-1).

This increase is best explained as part of the growth in special-interest politics, as opposed to the politics of party identification or patronage (McFarland 1976). Many kinds of single-issue groups formed or expanded their activities at the time, expressing concerns about peace, arms, schools, communism, poverty, consumer affairs, abortion, and electoral reform. Political participation increases with education, and the proportion of college-educated Americans rose sharply after World War II. By the 1960s, activists had models to emulate in popular attempts to rectify civil rights and end the war in Vietnam. The news media, especially television, provided a salient platform from which to press grievances. These factors encouraged activism on many fronts, including technological criticism.

Partly in response to this crescendo, and partly because of growing interest in the space program, medical innovations, and environmental issues, reporters began specializing in science, medicine, and the environment in the 1960s. The very presence of specialist reporters, who came to know and trust sources within the scientific and environmental commu-

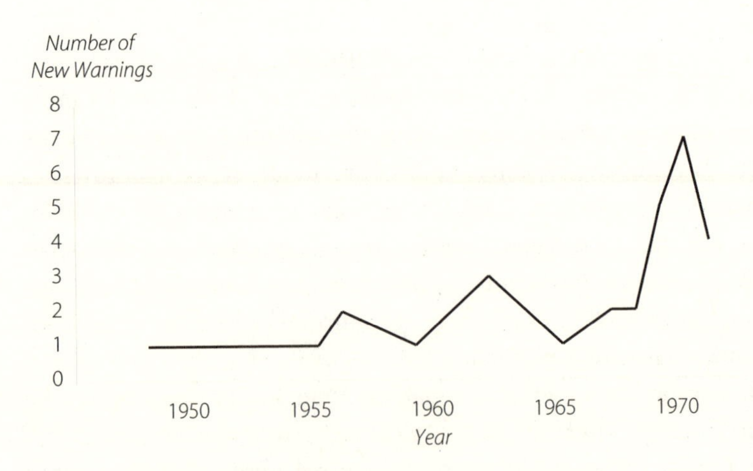

Figure 3-1. Number of New Warnings by Year (1948–1971)

nities, facilitated new warnings concerning pollutants, food additives, nuclear weapons, and consumer goods. Journalists joked about the hazard of the week. It was not the pace of technological change that made readers respond to *Future Shock* but the unprecedented amount of highly publicized controversy.

Lawless's sampling of warnings is impressive in its diversity. It omits some notable controversies of the period, however, including those over the atmospheric testing of nuclear weapons, the antiballistic missile (ABM), the supersonic transport (SST), and nuclear power plants.

To place the sample cases and other important warnings in historical context, I group them here into three unifying themes: pollution, food and drugs, and military technologies, with the latter emanating from protests over the war in Vietnam. These themes capture 30 of the 31 warnings Lawless studied. (The one case that does not fit is the warning that football injuries are more frequent on synthetic turf than on grass [Case 31, true].) As I trace the sources of these warnings and the links among them, I indicate in parentheses whether—from today's perspective—each warning proved true or false, based on criteria I set forth in Chapter 5.

WARNINGS ABOUT POLLUTION

The first set of warnings were both cause and effect of the greater environmental awareness that developed in post–World War II America. Conventional histories place the roots of modern environmentalism in the conservation movement of Theodore Roosevelt's day. The 1920s through the 1950s are seen as a hiatus during which little was accomplished until environmental activism reemerged in the 1960s. This view ignores important precursors unconnected to conservation. Urban problems—tuberculosis, waterborne diseases, sewage and horse manure in the streets, filthy air, spoiled food—were dealt with by public health officials, social reformers, sanitary and waterworks engineers. Improvements throughout the first half of the twentieth century provided city dwellers with safe food and water, removed bacteria-laden sewage, and eliminated scourge diseases. Modern environmentalism is as rooted in these urban problems, especially pollution, as in conservation.

The industrial effort of World War II and the suburban growth that followed the war badly contaminated the air and water with chemicals. Los Angeles was an extreme example, its mixture of sunshine, automotive exhaust, temperature inversions, smoke, and surrounding mountains producing opaque air commonly called "smog" or "smaze." Residents regarded this pollution as an acceptable nuisance, even a sign of prosperity. Smog was a subject for Hollywood comedians. Then, in 1948, the industrial town of Donora, Pennsylvania, surrounded by hills, suffered a combination of fog and factory emissions that caused 20 deaths and made nearly 6,000 people ill within a few days. Reported on the front-page of the *New York Times* of October 31 and November 1 and in *Life* magazine ("Death over Donora," Nov. 15, 1948: 107–110), Donora was the nation's first alarm about obvious pollution, although it evoked little sustained response (Case 16, true). The problem again became distressingly visible in 1969, when an accident at a drilling rig off the shore of prosperous Santa Barbara, California, triggered a release of oil that tarred beaches and killed wildlife. Angry citizens, with the help of journalists, woke the nation to the threat of pollution. The same year, Cleveland's Cuyahoga River caught fire, illuminating the problem of floating debris and oil.

The first Earth Day, April 22, 1970, was a consciousness-raising event proposed by Senator Gaylord Nelson, a Democrat from Wisconsin. Adapting the teach-ins, rallies, music, and speeches of 1960s movements, Earth Day was blessed with perfect spring weather across the nation, and was less confrontational than demonstrations for civil rights or against the war in Vietnam. Within a few months, Congress enacted the Clean Air Act of 1970, aimed primarily at improving health by reducing urban smog stemming from auto emissions. The Clean Water Act followed in 1977. Like the Clean Air Act, it strengthened earlier legislation.

More interesting for our story than warnings about blatant air and water pollution are those concerning trace pollutants. With the exception of crank warnings in the 1930s about food poisoned by then-new aluminum cookware, or contamination by germs, people had never before thought of tiny exposures to anything as threatening to health. I recall as a child in the 1940s shopping for shoes with my mother. To check the fit, she, the salesman, and I peered down into an x ray fluoroscope while I wiggled my toes in the shoes. A fluoroscope was also part of routine vis-

Twenty million people took part in Earth Day on April 22, 1970 (from http://www.aqmd.gov/monthly/jul97cov.html).

its to the pediatrician: the doctor and my mother would gaze at the glowing screen revealing the inside of my torso. The test had no important diagnostic value for healthy children, but people did not then regard low-level x rays as too risky for their entertainment benefit. In 1944 physicians shrunk my chronically infected tonsils with x rays. At that time, it was a progressive treatment given to thousands of children. We learned decades later that the practice caused thyroid cancer (Mazur 1981).

DDT was introduced as a body louse powder during World War II, and it successfully controlled malaria, typhus, and other insect-borne diseases among the troops. After the war its application in developing countries saved millions of lives by limiting malaria, an unprecedented victory over disease that earned its discoverer, Paul Müller, the 1948 Nobel Prize for Medicine. Civilians bought DDT "bug bombs" to kill insects around the house, while trucks and airplanes sprayed DDT over fields and neighborhoods, often using far more than recommended.

By the 1970s, many Americans had become so fearful of chemicals and radiation that cynical commentators would later call the nation "chemophobic." Americans had apocalyptic images of trace poisons insidiously

"Thank modern science for making it so easy it's fun to kill insects when they move in to live with you..... you see a misty white spray. Quickly it dissapears, but invisible atoms of safe Pyrethrum, in effective combination with DDT, are loose in the room.... More and more women tell us the only thing better than a Westinghouse Bug Bomb is two of them—one upstairs, always handy to bedrooms... one downstairs, handy to kitchen, living and dining rooms..."

This advertisement for a "bug bomb," in the July 1947 issue of Good Housekeeping, *is typical of many ads appearing that summer for DDT insect sprays.*

seeping through the environment and into their bodies, causing an epidemic of cancer (Efron 1984; Whelan 1993). This was a remarkable turnaround from the relative lack of concern of the 1950s and the innocent awe of the prewar era. What caused this change?

The first popular warning about a trace poison was the protest against fluoridation of community drinking water, which began about 1950 (Case 8, false; see Mazur 2001). Fluoridation has its roots in the 1930s, with reports that people living where water naturally contained fluoride had teeth that, while often discolored, were relatively free of cavities. Research showed that the benefit of cavity prevention could be obtained with little discoloration if the concentration of fluoride was as low as one part per million. In 1945 the U.S. Public Health Service supported the experimental addition of fluoride at this concentration to the drinking water of a few cities, intending to compare their cavity rates with those of control cities over the following 10 years. Some Wisconsin dentists, enthusiastic over the low cavity rates reported during the first years of the study, urged that mass fluoridation be promoted immediately. The Public Health Service first resisted, saying that it would wait until completion of the 10-year experiment. In 1950, however, the Public Health Service recommended fluoridation across the nation. By 1951 the American Dental Association and the American Medical Association had added their endorsements.

Almost immediately politically conservative citizen groups in Wisconsin protested adding a toxic chemical to their drinking water, arguing that fluoride (in higher doses) was a rat poison and that involuntary fluoridation amounted to mass medication, a step toward socialism. The movement spread across the United States, gaining strength from concerns over the federal government's susceptibility to communist influences. When communities held referenda on fluoridation, the measure was usually voted down (Mazur 1981).

In the mid-1950s the nation became concerned with another trace pollutant: radioactive fallout from nuclear weapons testing. Americans applauded the atomic bombings of Japan, which brought a quick end to the war, and believed the weapons were in the right hands. That comfort changed surprisingly quickly after the Soviet Union exploded its first atomic bomb in 1949 and then detonated a hydrogen device in 1953.

The Cold War reinforced a sharp left–right polarization in U.S. politics, with each side seeking its own symbols and issues with which to wage the debate. In 1954 a Japanese fishing boat, the *Lucky Dragon*, was accidentally showered with fallout from an American hydrogen bomb test. The incident precipitated first in Japan and then in the United States and Europe a leftist movement aimed specifically at halting atmospheric testing. This anti-testing movement—and more generally opposition to the arms race and nuclear confrontation with the Soviets—had an enormous effect on public perceptions of environmental radiation and its dangers.

Environmental radiation became a presidential election issue in 1956. In that year, partly in response to concern about radiation, and anticipating the introduction of peaceful nuclear energy, the National Academy of Sciences released a highly publicized report showing that the U.S. population was exposed to far less radiation from weapons-testing fallout than from naturally occurring sources such as cosmic rays and, surprisingly, from medical and dental x rays (Case 14, true; see Mazur 2000). The practice of illuminating children's feet with fluoroscopes was held up as a particularly foolish exposure—worse than fallout and without benefit (Case 13, true; see Mazur 2000).

Scientists who opposed atmospheric testing were not convinced that it was relatively innocuous. Linus Pauling, a Nobel Prize–winning chemist and influential advocate, warned that fission products from nuclear explosions, especially strontium 90, descend as radioactive precipitation, contaminating grass. The radioactive agents are passed on to children in the milk of cows that have eaten the grass. "There exists a real possibility," he claimed, "that the lives of 100,000 people now living are sacrificed by each bomb test or series of bomb tests in which the fission products of 10 megatons equivalent of fission are released into the atmosphere" (Pauling 1958, 108).

The fallout controversy reached a sudden resolution in 1963. The year before, President John F. Kennedy and Soviet General Secretary Nikita Khrushchev had brought their nations close to war in the crisis over missiles in Cuba. Afterward, in relief, the two leaders agreed on a series of confidence-building measures, including a mutual ban on atmospheric testing. Nuclear weapons were thereafter tested underground.

Other warnings about radiation emerged. Some were minor, like the notice in 1967 that improperly shielded color televisions were emitting x rays (Case 15, true). Others, like the controversy over nuclear power, were major.

Power plants were the peaceful side of atomic energy, promising to generate electricity "too cheap to meter." The first citizen protest against a nuclear power plant, an experimental breeder reactor in Detroit, occurred in 1956. The central concern was the possibility of an accident that would breach the reactor and release radiation over the city. During the early 1960s, local groups intervened against a few more plants, partly over accident potential but increasingly for fear of the low-level radiation emitted by a normally functioning nuclear plant. In 1970 radiation biologists John Gofman and Arthur Tamplin, of the University of California's Lawrence Radiation Laboratory, claimed that radiation from properly operating nuclear power plants would produce 32,000 additional cancers and leukemias per year. Experts representing the nuclear power industry vigorously denied the assertion.

By the first Earth Day in 1970, antinuclear groups around the nation were coalescing and publicizing a new hazard: spent reactor fuel, which remains radioactive for thousands of years. The industry was planning to permanently dispose of these wastes in supposedly impermeable underground salt formations. In 1970 the U.S. Atomic Energy Commission announced its tentative selection of salt beds near Lyons, Kansas, providing that small town with a dependable source of long-term employment. Opposition arose not from Lyons but from elsewhere in the state, with warnings that holes in salt beds would allow water to seep into stored waste, risking radioactive contamination of underground aquifers (Case 25, true). Amidst a national energy crisis precipitated by the Arab oil embargo of 1973, the nuclear power controversy ballooned into a massive protest, inflamed in 1979 by the accident at Three Mile Island, which marked the end of orders for new nuclear power plants in the United States.

The protesters of the 1950s against fluoridation and radioactive fallout usually occupied opposite ends of the political spectrum, yet their risk messages were essentially the same (Mazur 1981). Both groups objected to involuntary chronic exposure of large populations to low doses of agents that were known to be very dangerous at higher doses. Both sets of protesters regarded distant and misguided leaders of government and industry as responsible for placing populations at risk. Both accused these parties of ignoring accumulating scientific evidence of toxicity from chronic low-level exposure. Both envisioned the poisons emanating from technology as insidiously contaminating the purity of nature. Both emphasized that trace poisons become increasingly concentrated as they are consumed by species higher up the food chain. Both saw pollution as symptomatic of the moral decay of society. Both worried about cancer.

These elements constituted the ideology of Rachel Carson's *Silent Spring* (1962), perhaps the most influential book published in the United States since *Uncle Tom's Cabin*. Carson warned that DDT (Case 19, true) and other pesticides persist and accumulate in the environment, harming wildlife and perhaps causing cancer in humans. Her ideas about chemical carcinogenesis came mostly from Dr. Wilhelm Hueper of the National Cancer Institute, who blamed trace industrial chemicals for the cancer epidemic he believed the nation was experiencing. (At the same time, he discounted cigarettes as an important cause of lung cancer.) It is ultimately Hueper, through Carson, who should be credited or blamed for America's preoccupation at that time with pollution-induced cancers. Carson's own illness may have influenced her view. Diagnosed with breast cancer in 1957, she had a radical mastectomy in 1960 and continued to be treated for the disease until her death in 1964.

Carson's critics charged that she was unscientific and inaccurate. Yet Carson held a master's degree in biology and was a highly competent science writer who had worked for years as an editor with the U.S. Fish and Wildlife Service. *Silent Spring* contains few factual errors, given the scientific information of the day (Marco, Hallingworth, and Durham 1987). Her book may be faulted on other points. Production of synthetic chemicals, which greatly increased in the twentieth century, cannot account for a large portion of human cancers, because there has not been a large increase in the age-adjusted cancer rate, after smoking and improvements in diagnosis are taken into account (Howe et al. 2001). Excessive use of DDT did hurt certain bird populations, especially raptors, but Carson was incorrect in predicting that the robin—the symbol of spring—would become extinct. Still, her claim that pesticides were overused, polluting the environment to a degree that damaged wildlife and possibly threatened human health, was correct and timely.

Why did *Silent Spring* have so large an impact in convincing Americans of the hazard of trace pollutants, when so many were unconvinced of the danger of smoking a pack a day of cigarettes? Carson's skill as a writer was helpful, and the book also appeared shortly before publication in 1962 as three installments in *The New Yorker*. Much of the magazine's readership at that time would not have known a chlorinated hydrocarbon from a pileated woodpecker, but interest in the fallout controversy was high and pesticides appeared as a corollary issue. In *Silent Spring,* Carson repeatedly compares the dangers of pesticide contamination with radioactive fallout, most effectively in her "Fable for Tomorrow" at the opening, in which a town dies because of a mysterious white powder that has fallen like snow from the skies.

Just when Carson's first installment appeared in *The New Yorker* (June 1962), the *Washington Post* ran a front-page story (July 15) about how Dr. Frances Kelsey of the U.S. Food and Drug Administration (FDA) had prevented the agency from approving the tranquilizer thalidomide, saving the United States from the tragedy of armless and legless babies that had occurred in Europe and elsewhere (Case 10, true). Kelsey instantly became a national heroine. The validation of her suspicion about thalidomide greatly boosted media attention to Carson's newly publicized warning about DDT, seemingly enhancing its credibility.

Attempts by the chemical industry to discredit Carson only focused more attention on her warning. Eventually, in 1972, the United States banned DDT, an action emulated by some developing countries, which quickly suffered increases in malaria. Other federal actions spurred by *Silent Spring,* and more widely applauded, were the 1976 Toxic Substances Control Act, requiring industrial chemicals to be tested for toxicity, and the Resources Conservation and Recovery Act, governing the disposal of hazardous waste.

By the 1970s other low-level pollutants were identified as hazardous. One was asbestos, which some readers may remember from childhood as the "friendly mineral" that insulated our schools and protected us from fire. By the mid-1970s, asbestos had been shown to cause a rare cancer, mesothelioma, in workers who breathed the fiber (Case 20, true). More industrial effluents and emissions became suspect, as when the Reserve Mining Company, which for years had polluted Lake Superior with its taconite (iron ore) tailings, was accused in 1969 of endangering drinking water with taconite (Case 21, false). This suspicion became more plausible in 1973, when asbestos was discovered in the tailings.

In 1970 Lake Erie fish were found with high mercury levels traceable to manufacturing plants on the Canadian and U.S. shorelines (Case 17, true). Lake Erie was already suffering badly from eutrophication, a major degradation in water quality caused by nutrients in sewage and farm runoff. While not a human health problem, the process of eutrophication leads to excessive growth of algae, which depletes the water of oxygen, causing fish kills, algae dieback, and foul odors.

Laundry detergents were also fingered as culprits in water pollution, both for their phosphates and their foaming action on top of the water. In 1970 detergent companies substituted nitrilotriacetic acid (NTA) for phosphates and introduced enzyme detergents with improved stain-removing qualities. Both additives were accused of being health hazards (Cases 22 and 23, false).

WARNINGS ABOUT FOOD AND DRUGS

Jonas Salk's killed-virus polio vaccine was a triumph of medical technology—the start of a successful drive to eliminate a dreaded child crippler that reappeared every summer. The most hyped development of postwar medicine, it was intensely promoted by the National Foundation for Infantile Paralysis, a private organization that was the creator of the March of Dimes, a charity that addressed polio and later other children's disabilities (Wilson 1963). So great was public interest in the new vaccine's field trial that the foundation's announcement in April 1955 of a successful result was broadcast live on television, reaching medical researchers, journalists, and public viewers at the same time.

At first, there was euphoria and an immediate drive to vaccinate the nation's children in time for the summer season. Then, suddenly, newspapers reported cases where children contracted polio shortly after being vaccinated (Case 9, true). Many parents were frightened away from the program. One of the companies manufacturing the vaccine had failed to kill the virus completely, causing some children to become infected. This episode badly damaged the credibility of the FDA, which was supposed to ensure the vaccine's safety.

The FDA was a frequent participant in postwar technical controversies, sometimes issuing warnings, sometimes denying warnings from elsewhere. The agency had basked in public admiration in 1962, when its Dr. Kelsey prevented approval of thalidomide, but the FDA often found itself in a defensive posture, accused of errors, delays, or exaggerations about a product's safety. Many of these problems arose from the growing perception of trace contaminants as seriously harmful, producing regulatory pressures the agency had never experienced before.

The FDA originated with the 1906 Food and Drugs Act. Intended to prevent adulteration and spoilage of food and to eliminate poisonous chemical preservatives in processed food, the act was a product of the reformist movements at the beginning of the century. A catalyst was Upton Sinclair's *The Jungle* (1906), a novel describing filthy conditions in Chicago's meatpacking plants. Drugs were less of a concern than food, the major worry being fraud from quack nostrums.

Early on the agency verified the safety of various substances by feeding them to a regularly assembled group of volunteers known as "the poison squad." This method was not as foolhardy as it may now seem, because medicines of that time were relatively powerless to either treat or harm patients. Microorganisms such as botulin were known to be lethal in very small doses, but chemicals of comparable potency that might serve as drugs or poisons were unknown. Chronic exposure to some heavy metals was recognized from workplace experience to be dangerous, but these exposures involved substantial cumulative doses. (Mercury poisoning, causing slurred speech, irritability, and memory loss, was an occupational hazard of felt workers, giving rise to "mad hatter's disease" immortalized by Lewis Carroll.) There was no great concern about using lead or arsenic pesticides on food crops as late as the 1940s, when DDT replaced them.

The FDA's attention was directed toward trace hazards in 1958, when Congress added the Delaney Clause to the Food, Drug, and Cosmetics Act. This clause banned the addition to processed food of even a trace of any chemical shown to cause cancer in animals fed very large amounts. This first and still most absolute legal protection against trace poisons carries the name of a conservative Democrat from working-class Queens, New York. First elected to the House of Representatives in 1945, James Delaney had no interest in chemicals until a colleague suggested that pesticide contamination of food was ripe for investigation. The earliest critics of DDT had just begun to express their concern over its rapidly increasing and often careless use. Delaney convinced the House speaker to create a special committee to investigate, with Delaney as chair. The committee held hearings from 1950 through 1952, calling experts with diverse views on the use of chemicals, including Wilhelm Hueper, who had educated Rachel Carson about chemical carcinogenesis. The committee gave considerable attention to DDT, and after the Public Health Service endorsed fluoridation in 1951, to fluoridation as well. Delaney became a strong opponent of fluoridation, calling it "an unnecessary health risk and unwarranted intrusion on the rights of our citizens" (Delaney 1975, 23,729). Delaney's clause did not follow immediately from these hearings but was enacted in 1958, after congressional interest in chemicals had grown.

A stream of warnings flowed from Rep. Delaney's clause and the concerns leading to it. The first came in 1959, 17 days before Thanksgiving. Arthur Flemming, secretary of the Department of Health, Education, and Welfare (at the time the parent department of the FDA), announced at a press conference that part of the cranberry crop grown in Washington and Oregon had been contaminated with an herbicide that had produced can-

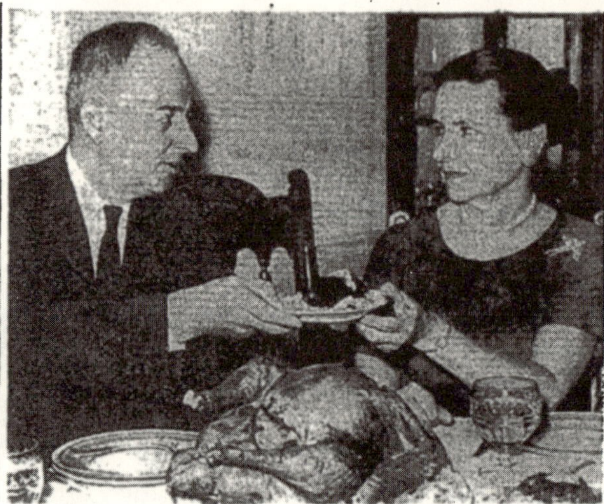

The New York

NEW YORK, FRIDAY, NOVEMBER 27, 1959.

Associated Press Wirephoto

'MORE CRANBERRIES?' Arthur S. Flemming, Secretary of Health, Education and Welfare, is offered cranberries by his wife at Thanksgiving dinner in their Washington home. He recently charged that some of the cranberry crop was tainted by weed killer.

KRISHNA MENON DISPUTES CRITICS OF DEFENSE PLAN

He Denies Belittling Chinese Threat to India's Border —Nehru Supports Him

By PAUL GRIMES

Special to The New York Times.

NEW DELHI, India, Nov. 26 —Defense Minister V. K. Krishna Menon strongly defended himself today against charges that he had underestimated the dangers in Communist Chinese aggression.

"When the time comes that I have to carry a card of patriotism," he told Parliament, "it will not be worth carrying."

It was Mr. Krishna Menon's first public reply to mounting criticism here of himself and India's defense policies. Many widely respected Indians have demanded that he resign.

For twenty minutes the Defense Minister stood in the lower house and sternly insisted

This photo appeared on the front page of The New York Times of November 27, 1959, captioned, "'MORE CRANBERRIES?' Arthur S. Flemming, Secretary of Health, Education and Welfare, [...] recently charged that some of the cranberry crop was tainted by weed killer." Flemming was attempting to restore the badly damaged reputation of the cranberry crop.

cerous growths in rats. Because consumers could not know where their store-bought cranberries had been grown, Flemming advised against purchasing cranberries, fresh or canned (Case 2, false). Following Flemming's interpretation of the new Delaney Clause, the entire cranberry crop was recalled pending tests by the FDA for contamination. At that time the cranberry market was almost entirely seasonal, and growers strenuously objected, fearing that their entire year had been ruined. Press coverage was enormous. Fortunately, cleared berries were allowed back on the market in time for Thanksgiving.

Two weeks later Secretary Flemming again made headlines, this time announcing that the use of diethylstilbestrol (DES) to fatten poultry would be halted and that the government would buy and remove from the market millions of treated chickens. A synthetic estrogen, DES had been approved by the FDA in the 1940s as a human drug, and by the Department of Agriculture for use in poultry farming. A pellet of DES, implanted in a young rooster's head, slowly dissolved to act as a chemical castrator, producing a less muscular bird with plump flesh for the dinner table. Any unused DES would be removed with the head at slaughter, and it was assumed that chicken pieces sold in the store would be free of the chemical. In 1957, Delaney inserted into the *Congressional Record* a charge that DES was a carcinogen (Case 3, true). The FDA promptly denied this, but the warning nonetheless reached the *New York Times.* By late 1959, DES had produced cancers in test animals, and improved analytical techniques could measure residues as low as 20 parts per billion in pieces of chicken. Flemming again said that the Delaney Clause required a ban.

DES was by then being used to fatten sheep and cattle, a practice allowed to continue because residues had not been found in their meat. Cattle were increasing treated this way until 1971, when Dr. Arthur Herbst linked rare vaginal cancers in young women to their mothers' use of DES, years before, as a drug to prevent miscarriage. Though it does not follow from this high-dosage effect that trace residues in meat were necessarily dangerous, as a precaution these uses of DES were banned.

The FDA in the 1960s seems to have become hypercautious in regulating drugs as well as food. In 1962 the agency banned the sale of fish protein concentrate, a nutritious and low-cost food supplement prepared from whole fish that was considered by some to help the world's hunger problems. To keep production costs low, fish were processed whole, including head and guts, which the FDA regarded as unsafe and unwholesome for consumers (Case 7, false). A study by the National Academy of Sciences concluded that the FDA had reacted too strongly, and in 1967 the ban was lifted. However, the product never enjoyed a large market in the United States.

Frances Kelsey, who had prevented U.S. approval of thalidomide, was placed in charge of the FDA's oversight of a potential new wonder drug, dimethyl sulfoxide (DMSO), which seemed useful in treating burns, arthritis, headaches, and sprains, as well as in serving as a vehicle for carrying other drugs into the body through the skin. DMSO has the remarkable property of being tasted in the mouth very soon after being rubbed on the skin, signaling the speed with which the body absorbs it. Obtained from wood pulp, DMSO can be cheaply purchased by anyone. This led to considerable unregulated experimentation, eliciting a warning in early 1965 that it might be dangerous to humans (Case 12, false). With the observation in 1965 that DMSO can cause eye damage in experimental animals, the FDA banned clinical tests on humans. Proponents objected with some justice that a cheap and beneficial drug was being suppressed (McGrady 1973). The FDA gradually relaxed its restrictions. In 1973 a study by the National Academy of Sciences concluded that DMSO did offer some benefits, and that eye damage or other adverse effects of normal use had not been established. Today DMSO is approved as a prescription medication for bladder problems, although its original promoters offer it on the Web for other uses.

The pace at which warnings appeared in the news noticeably accelerated in the late 1960s. The FDA had approved the first birth control pill in 1960. In 1967 came the first warning that women using "the Pill" suffered increased health risks (Case 1, true). Although newer pills with different and lower hormone doses have reduced these risks, the issue remains contentious today. The Pill had important social effects, with its use linked to both the sexual revolution and the women's movement.

The media played a role in marketing both the Pill itself and an ideal built around the Pill—a woman who was confident, sexually sophisticated, and slender. This ideal helped manufacturers market a variety of products, including cyclamate-sweetened diet drinks. Consumption of the drinks skyrocketed until 1968, when the FDA concluded, partly on the basis of internal studies, that unrestricted use of cyclamate was unwise (Case 4, false; see Mazur and Jacobson 1999). This became a contentious issue within the agency. A mid-level FDA scientist, Dr. Jacqueline Verrett, appeared on network television news showing thalidomide-like birth defects in chick embryos that were produced when cyclamate was injected into eggs. Days later Robert Finch, secretary of Health, Education, and Welfare, announced that cyclamate had caused cancer in rats and therefore, under Delaney, would be banned. The diet soda industry quickly responded by substituting a saccharin and sugar mixture, but in 1977 the FDA proposed to ban saccharin when it, too, was found to cause cancer in rats. Congress quickly set the ban aside, and saccharin continues in use

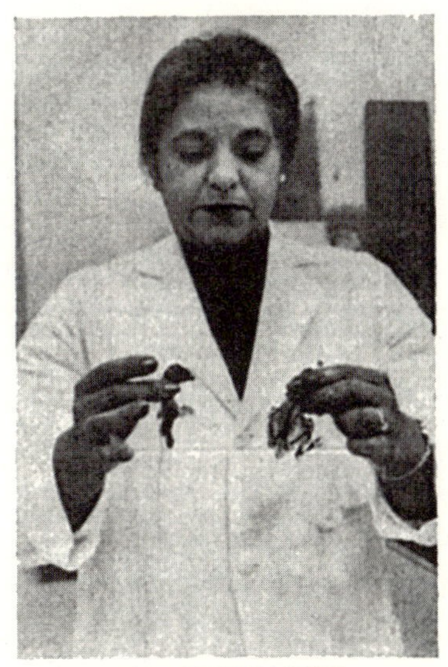

DR. JACQUELINE VERRETT
Not for chicks, not for people.

Time *magazine of October 17, 1969, showed Dr. Jacqueline Verrett, an FDA research scientist, with chick embryos deformed by exposure to cyclamate. Photograph by Walter Bennett.*

today in the United States. Canada takes the opposite regulatory posture, placing more stringent restrictions on saccharin than on cyclamate.

Within days of the cyclamate ban, consumer advocate Ralph Nader warned Congress that monosodium glutamate (MSG), an additive in baby food, was harmful to infants (Case 5, false). Long common in Asian food, MSG's derogation had begun the year before in the prestigious *New England Journal of Medicine* (April 4, May 16, and July 11, 1968), with numerous letters from physicians suggesting it as the cause of "Chinese restaurant syndrome," a set of allergic reactions suffered by some patrons of Asian restaurants. With the suggestion that MSG was also harmful to infants, public concern greatly increased. Baby food manufacturers removed MSG from their products, where it had served only to satisfy the palates of mothers.

In 1970 the FDA recalled 1 million cans of tuna because of mercury contamination (Case 18, false). This warning was nearly simultaneous with, and derives from, the 1970 warning about mercury pollution of Lake Erie from shoreline factories. The tuna story took an unexpected turn with the discovery that museum fish specimens, predating the era of industrial pollution, had similar levels of mercury. Shortly afterward, most fish were allowed on the market. Although the mercury warning of 1970 was overblown, the FDA today cautions women who are or may become pregnant, or who are nursing, not to eat fish high in methyl mercury, particularly shark, swordfish, king, mackerel, and tilefish.

Lawless includes two other food and drug warnings, both from 1971. Infants at that time were often bathed with a disinfectant called pHisoHex, until the announcement that its active ingredient, hexachlorophene (HCP), can be absorbed through infant skin in toxic amounts (Case 11, true). Today pHisoHex (the brand name for the most common preparation of HCP) is used in hospitals but not on babies.

Finally, we have a case of old-fashioned botulism. Once associated with improperly canned food, botulism is rare today because of improvements in food processing. Still, it does occur, as it did in New York when a banker died and his wife became ill from an infected can of Bon Vivant vichyssoise soup (Case 6, true). A prompt warning from the New York Health Department led to a nationwide attempt to remove all of Bon Vivant's soups from stores. The soup company went bankrupt a few months later.

WARNINGS ABOUT MILITARY TECHNOLOGY AND THE WAR IN VIETNAM

Environmentalism was growing in the 1960s, but it did not occupy center stage. The real action during that eventful decade was in leftist politics of the counterculture, first as part of the civil rights movement, later relating to protests against the war in Vietnam. A third cluster of Lawless's warnings, related to military technologies, inevitably reflected widespread protest over the war.

At first U.S. involvement in Vietnam evoked little domestic interest. Minor military action under President Kennedy was acceptable to Americans. The turning point came after Kennedy's successor, Lyndon Johnson, beat the hawkish Barry Goldwater in the presidential election of 1964. During the campaign Goldwater had advocated an extreme escalation of the war; Johnson spoke moderately but disingenuously. After the election Johnson escalated the war, inciting by 1965 a vehement protest movement. The conservative writer William F. Buckley remarked, "They told me if I voted for Barry Goldwater we would be bombing North Vietnam, so I voted for Goldwater and sure enough, we are bombing North Vietnam."

The 1968 Democratic convention, with confrontations between Chicago police and civil rights and Vietnam War protestors, was the most violent in memory. Chemical Mace sprays, introduced a few years earlier for riot control, were thought by some activists to cause permanent injury (Case 30, false). The Mace controversy peaked after its use at the convention.

In the mid-1960s aerospace optimists proposed a system to shoot down incoming Soviet warheads with antiballistic missiles (ABMs), a technically difficult task often portrayed as hitting a bullet with a bullet. As late as 1967, President Johnson and Secretary of Defense Robert McNamara opposed deployment of an ABM system but then reversed their decision, providing antiwar liberals in Congress and universities with a new issue on which to challenge the administration. Opposition to ABMs became strongly tied to opposition to the war.

By 1969, when Richard Nixon succeeded Johnson, the Vietnam War protest was expanding to all things military, including university research funded by the Defense Department and campus ROTC. The ABM controversy escalated, with both sides recruiting scientists in a battle of petitions in the *New York Times*. Proponents thought the system technically feasible; opponents said it was not.

By 1970 the supersonic transport (SST), supported by the Nixon administration, had become another focus of dispute. A civilian airliner, the SST was nonetheless close enough to military technology to attract largely the same proponents and opponents who had aligned on ABMs and the war (Mazur 1981). The main technical issue concerned the possibility that nitrogen oxides injected by a fleet of high-flying SSTs might deplete stratospheric ozone, a worrisome consequence that in 1985 would reemerge in association with chlorofluorocarbons.

The warnings on military technology that Lawless studied appeared between 1968 and 1971, a period of intense protest over the war in Vietnam. The Army triggered the first of these warnings when a 1968 test at its Dugway Proving Ground in Utah accidentally released poisonous gas, killing 6,000 sheep. At first the Army denied any release, but suspicions immediately focused on Dugway and its potential threat to human life (Case 27, true).

The Dugway incident inspired protests a year later at an Atomic Energy Commission plant at Rocky Flats, Colorado, run by Dow Chemical Company to produce plutonium for nuclear weapons. The plant suffered a disastrous fire in May 1969, which released very little radioactivity but raised concerns over whether its normal operations were causing contamination that endangered the health of residents (Case 24, true).

The protest at Rocky Flats was led by a group, composed mostly of physicists, calling itself the Colorado Committee for Environmental Information (CCEI), part of a network inspired by ecologist Barry Commoner to provide scientific information relevant to public issues. CCEI began in 1967 as an academic seminar at the University of Colorado but switched to activism when members became concerned about the Dugway sheep kill. The group produced a report on similar danger at nearby Rocky Mountain Arsenal and investigated a 1969 fire at the Dow plutonium plant (Nelson 1970).

Also in 1969, Wisconsin residents learned of Navy plans to place in their state 21,000 square miles of underground cables, intended as an antennae for radio waves of extremely low frequency (ELF). ELF transmissions follow the curvature of the earth, allowing communication with U.S. submarines anywhere in the world. A coalition of antiwar and environmental activists and local farmers raised the first warning that ELF radiation, although well below the energy of x rays, might cause health problems in humans (Case 29, false). By 1972 the Navy had reduced the size of its planned grid and dropped Wisconsin as a site. The project continued to be scaled back, never finding a site free of opposition. It seems finally to have been abandoned.

Earth Day 1970 came at a time of growing opposition to the war and considerable public concern over chemical warfare. In November 1969 President Nixon curbed use of chemical weapons and banned germ weapons. An international treaty was discussed to eliminate all these weapons. Yet a year after Dugway, an accidental release of nerve gas on Okinawa hospitalized 23 American soldiers and 1 civilian. The first Earth Day came amid protests in Washington State and Oregon against shipment of the chemical weapons from Okinawa to the United States. Three months later, Rep. Cornelius Gallagher, a Democrat from New Jersey, told the press that the Army planned to ship two trainloads of poison gas weapons from two southern states to New Jersey for dumping in the Atlantic, raising the possibility of an accidental release during transport (Case 28, true). The media reported the ensuing controversy and the voyage of the weapons to their ocean burial in detail.

The last case involving the technology of war recalls the controversies in the 1950s over nuclear fallout. After the 1963 agreement signed by Kennedy and Khrushchev, U.S. and Soviet nuclear weapons testing occurred underground. In mid-April 1971 the Atomic Energy Commission announced the planned underground detonation of a hydrogen bomb, part of the ABM developmental program, late in the year on Amchitka Island, a small piece of land off the Alaskan coast, and the site of earlier tests. By the end of April, opposition had been voiced in the Alaska legislature and by the Federation of American Scientists, a nonprofit organization especially concerned about nuclear weapons. Warnings were raised about the possibility of setting off a disastrous earthquake or tsunami (Case 26, false). Protests were wide-ranging, occurring in the courtroom, in the press, and at sea. The Audubon Society urged the Atomic Energy Commission to cancel the test lest it damage wildlife. Identifying themselves as a new group called Greenpeace, activists aboard an old fishing vessel crossed the North Pacific in protest. The hydrogen bomb exploded without causing an earthquake, tsunami, or other cataclysmic event.

THE INTERACTIONS AMONG WARNINGS

My designation of three themes—pollution, food and drugs, and military technologies—is a convenience, a device to group 30 cases (of 31) into a few sensible clusters. Other schemes might work as well, but there are limits if we require overarching themes that reflect the social reality of the period, as I believe that these do. For example, Ralph Nader's warnings could be considered as a separate cross-cutting theme. Nader became famous after his book *Unsafe at Any Speed* (1965) exposed General

Motors' Corvair as accident prone and the company as venal. Soon Nader was a familiar news source, often appearing before Congress and in other arenas, sometimes leading an alarm, sometimes lending support to other critics (McCarry 1972). Lawless specifically cites Nader among the critics of DES, MSG, fluoridation, HCP, x rays from defective televisions, shoe-fitting fluoroscopes, mercury in tuna, asbestos, and enzyme detergents—an incomplete list. In the 1970s, Nader became a major critic of nuclear power plants; his activism on many fronts continues to this day.

However one finally defines the integrating themes, it is inescapable that these public warnings did not arise in isolation. Nearly every one is connected to some other warning or broader concern, recently or currently in the news. In motivating partisans to support or oppose it, a technology's association with other contentious issues in politics and society may be as important as its intrinsic risk.

For example, warnings about specific military or military-like technologies raised between 1965 and 1971 came largely from people opposed to the war in Vietnam, and one warning often influenced another. The FDA was a central player in nearly all warnings related to food and drugs, sometimes facing the same interlocutors in one case after another. In Congress the same principals promoted various pieces of legislation to reduce pollution in the interest of health, often opposed by the same corporate interests. We also see connections between the technical positions taken by disputing experts and their policy positions on other issues, as we shall discuss in the next chapter.

CHAPTER 4

WHY EXPERTS IN TECHNICAL CONTROVERSIES DISAGREE

Virtually any case study of a technological controversy, including those Lawless studied, will contain references to each side's experts—the properly credentialed scientists, engineers, and physicians who buttress positions with technical information and undermine the scientific credibility of the other side. Laypeople are often confused and dismayed when one scientist contradicts another's facts. But science does not equal truth, and competent scientists have ample opportunity to disagree with one another, particularly on questions pushing the state of the art.

Alvin Weinberg (1972) has labeled some of these unresolved questions "trans-scientific" because they are, in practice, beyond the capacity of science to answer. One example is the disagreement over health effects of very low-level radiation. Obtaining significant results on the biological effects of very low-level radiation would require so many mice, so much time, so many scientists, and so much money and equipment that the experiments will probably never be undertaken.

Scientific disputes over the harmful effects of low-level radiation became prominent during the 1950s in connection with the warning about fallout from nuclear weapons tests, and resurfaced in the 1960s and 1970s during controversy over nuclear power plants. Warnings in the late 1960s came from Dr. John Gofman and Dr. Arthur Tamplin, research associates at the Lawrence Radiation Laboratory. Gofman was also professor of medical physics at the University of California, Berkeley, and a former associate director of Lawrence. In 1969 Gofman and Tamplin claimed that if the population of the United States were exposed to the maximum level of radiation permitted by federal standards, an additional 16,000–32,000 cases of cancer would occur each year. They recommended a 10-fold reduction in the federal standards (Mazur 1981).

The scientific reception of their claims was described in the prestigious journal *Science*: "[T]he Atomic Energy Commission (AEC) challenged their assumptions, disputed their estimates, and disagreed with their recommendation" (Boffey 1970, 838). While many experts disagreed with them, the two were reputable scientists whose arguments were not clearly wrong. As *Science* noted, "Most scientists who have worked on setting [radiation] standards believe that many of the assumptions made by Gofman and Tamplin are unjustifiable but find it difficult to disprove specific points" (Holcomb 1970, 854).

The radiation debate can be compared with the contemporaneous controversy over adding fluoride to community drinking water to reduce the incidence of dental cavities. On their face, the two controversies have little in common, but on closer inspection they are remarkably similar. Both focus on a technical question: what are the harmful effects, if any, of long-term exposure to low levels of either fluoride or radiation? Both fluoride and radiation are known to be lethal at high doses, and although there is no clear evidence of harmful effects in very low doses, neither, some experts argue, is there compelling evidence to the contrary.

Popular stereotypes portray the antifluoridationist as a kook, bigot, or extreme right-winger. While some opponents of fluoridation could reasonably be described in these terms, opponents have also included respectable scientists and physicians who feared possible toxic effects. Few neutral commentators have given serious consideration to their arguments. Instead, psychologists regarded opposition to fluoridation as an "antiscientific attitude" (Mausner and Mausner 1955). Social scientists, assuming that an informed voter could not rationally oppose fluoridation, referred to its frequent defeat in referenda as "democracy gone astray" (Crain et al. 1969). In contrast, the critics of radiation levels were given a respectful, if not hospitable, reception.

There are several plausible explanations for the different levels of respect accorded these two warnings. The radiation argument may have been more objectively sound than the fluoridation argument. Surely the scientists who warned against radiation—not only Gofman and Tamplin but also Nobel laureate Linus Pauling—had higher professional stature than those opposing fluoridation. Perhaps most important, the antifluoridation movement was originally associated with the "red-baiting" anticommunism of the late Senator Joseph McCarthy, which was anathema to the academic and scientific communities. Such political elements may be as important as scientific substance in determining how experts evaluate a warning. This chapter examines the reasons why scientists disagree and the techniques they use to present their arguments. I use the controversies over the effects of low levels of radiation and fluoride to illustrate the underlying sources of these disagreements.

USING RHETORICAL DEVICES TO SUPPORT OR OPPOSE A WARNING

Rhetorical similarities in the arguments over radiation and fluoridation as they appeared in news articles, speeches, congressional hearings, and polemical literature can shed important light on how proponents make their case. Perhaps these rhetorical devices, more than conflicting substantive arguments, are the main source of scientific and public confusion.

Here are several passages by Gofman and Tamplin warning against the dangers of low-level radiation from nuclear power and questioning the credibility of its sponsoring agency, the AEC. Each is followed by a similar quotation concerning the dangers of fluoridation and the irresponsibility of its sponsoring agency, the Public Health Service (PHS) (Exner, Waldbott, and Rorty 1957; see Mazur 1981, 15–18, for detailed citations on the sources of the quotations).

Radiation: Even though the water effluent at the release point may make the water drinkable...the fish grown in such water...cannot be eaten in any quantity without grossly exceeding "tolerance levels."
Fluoridation: People living in fluoridated cities who eat a good deal of seafood and drink tea and beer may easily ingest a combined fluoride intake far beyond...the tolerance limits.

Radiation: The AEC clearly demonstrated that when the chips are down on questions of protecting human beings and their environment, the promotional huckster role wins out handily over the public protector role.
Fluoridation: [Consider] the reckless arrogance, obstinacy, and unscrupulousness of the PHS in continuing to promote the program while ignoring and, where possible, suppressing evidence that it is neither safe nor genuinely efficacious.

Radiation: Where unknowns exist, always err on the side of protecting the public health.
Fluoridation: The public should have the benefit of the doubt and the procedure should be considered harmful until proved otherwise.

Radiation: Tamplin and Gofman presented evidence...that our allowable radiation exposures...are grossly unsafe.... The AEC response: Derision, denial, slander—but no evidence in refutation.
Fluoridation: [Critiques] of the proponent scientific data have been presented to the PHS.... Instead of dealing with subject matter itself, they attempt to show that the author is not qualified to discuss matters related to fluoridation.

Radiation (referring to the strategy of the AEC): Tell a big lie and tell it again and again as widely as possible.
Fluoridation (referring to the strategy of the PHS): A colossal lie, if repeated often enough, will be accepted a truer than truth.

Many of these statements could be transposed from one controversy to the other by changing "radiation" to "fluoridation," or "Atomic Energy Commission" to "Public Health Service," or vice versa. Lest this similarity be considered a unique characteristic of the critics, here are comparable passages from proponents of nuclear power and fluoridation.

Radiation: Radiation is by far the best understood environmental hazard.
Fluoridation: Never before has a public health measure been subjected to such thorough scientific scrutiny.

Radiation: A number of national concerns have converged to make up what we call the nuclear controversy...[including] an increasing distrust of science.
Fluoridation: The opposition to fluoridation can be attributed to...current suspicion of scientists.

Radiation: The risk of nuclear power is very much lower than the risk of alternate power sources.
Fluoridation: The risk...[of] fluoridation of water appears to be small in comparison with the dental benefits.

Radiation: The rationale behind permitting the release of small quantities of radioactivity...[is] that the environment has been radioactive from natural causes since the beginning of time. All natural solids, liquids, and gases contain radioactivity in varying amounts.
Fluoridation: Foods very common in our diet...contain fluorine in amounts varying from 0.14 to 11.2 parts per million. Therefore, the addition of fluorine at approximately 1.0 part per million to the water is not introducing a new element.

The most common rhetorical device is a variation on "there is no evidence to show that . . ."

Radiation

- No evidence exists for such an effect [due to rate of radiation delivery] on cancer induction by radiation in man.
- At the present time no valid evidence, based upon scientific observation, has been brought forward to prove that natural sources of radiation have produced injury to man in any way.
- There are no experiments that show that the integrated low-level effect [of radiation] is higher than that of the same amount given at one time.

Fluoridation

- No evidence has been produced that one part per million of fluoride in drinking water has or will harm any living person.
- Surveys...of kidney disease in fluoride as compared to nonfluoride areas have not produced any evidence of harmful effects upon the kidneys.
- [The American Medical Association is] unaware of any evidence that fluoridation of community water supplies up to a concentration of one part per million would lead to structural changes in the bones or to an increase in the incidence of fractures.

- In the accumulated experience there is no evidence that the prolonged ingestion of drinking water with a mean concentration of fluorides below the level causing mottled enamel would have adverse physiological effects.

Sometimes an expert uses this device to argue both sides of a controversy. The first two of the above passages on radiation quote Gofman—the first in his role as a nuclear critic, the second 12 years earlier when he was a proponent of nuclear power. Not surprisingly, there are countermoves to the "no evidence" rhetoric:

> Witness a typical statement by Mr. Frederick Draeger of the Pacific Gas and Electric Company: "There is no evidence that 170 millirads is harmful, and any new plant will actually emit only an infinitesimal fraction of that amount." Apparently, Mr. Draeger hasn't the slightest comprehension of what his statement "no evidence" really means. "No evidence" here means no one has looked. (Gofman and Tamplin 1971).

We find the same mode of argument in the fluoridation debate. For example, when the American Medical Association (AMA) stated that it was "unaware of any evidence" that fluoridation would be harmful, an opponent responded: "[If the AMA] had actually considered the evidence instead of trustingly accepting what McClure said about the evidence, they would not have been unaware of the dangers of fluoridation" (Exner et al. 1957, 79–80).

ARGUING PAST ONE ANOTHER

Some observers, trying to place their fingers on the points of disagreement between two experts, have concluded that the two do not disagree at all but rather are arguing about different points. This failure to confront each other's arguments is clearly present in the dispute surrounding the Gofman–Tamplin analysis of the expected number of deaths from the nuclear power program.

Gofman calculated that approximately 32,000 additional cases of cancer would occur annually in the United States as a result of exposure to the maximum radiation level permitted by federal standards (Mazur 1981). Dr. Victor Bond of Brookhaven National Laboratory stated the opposing view, claiming that nuclear power would cause just 0.02 cancers a year—that is, virtually no increase (Bond 1970).

Table 4-1. Alternative Calculations of Annual Number of Cancer Cases Caused by Radiation

Estimator	*Risk (cases per million people per millirem) (1)*	*Dose per year (millirems) (2)*	*Population (millions) (3)*	*Cancer cases per year (1) x (2) x (3)*
Gofman	0.94	170	200	32,000
Bond	0.1	0.001	200	0.02

Note: rem = roentgen equivalent man.

Source: Mazur 1981.

Calculations by both sides are compared in Table 4-1. "Cancer cases per year" is the product of three factors: risk (measured in cancers per million people per millirem[1] of exposure), dose per year (measured in millirems), and the size of the U.S. population (then 200 million). Because Gofman and Bond assigned different values to risk and dose per year, they arrived at markedly different values for cancer cases per year. Their values for risk differ by an order of only 10 (0.94 versus 0.1; see the next section). Their values for dose per year differ by an order of 10^5 (170 versus 0.001). Gofman's value for the latter was based on the permissible (but not actually achieved) level of exposure. Bond's value was his estimate of actual average exposure to the population, which is much smaller than the permissible exposure. The two calculations are about two different things. Yet Bond concluded:

> Dr. Gofman's speculations that 32,000 additional cancers per year will result from radiation exposure of the public under current "standards" simply do not conform to reality. They are in fact *in error* by a considerable margin for the present and for the foreseeable future. His figures have essentially zero validity in the context of power reactors. In this context, an upper limit estimate of the *correct* figure is well below one death per year in the entire U.S.A. (Bond 1970, 12; emphasis added).

Of course, the question of what is "in error" and what is "correct" depends on what is being calculated.

In both controversies there are similar patterns of conflicting contentions based on different premises. One of these involves the difference between acute and chronic forms of radiation or fluoridation poisoning. Proponents of fluoridation and nuclear power occasionally argued for the safety of their proposed technology by indicating how difficult it would be for a person to receive the relatively high dose associated with acute poisoning: "Even at one part of fluoride per million parts of water, to get a lethal dose from it you would have to drink 400 gallons at one sitting" (Forsyth 1952, 1504).

Critics, however, were concerned with chronic poisoning, which is associated with much lower dosages: "All the talk about the hundreds of gallons [of water] you would have to drink at one time to get sick refers to acute [fluoride] poisoning, which isn't even under consideration.... What is important is that the presence of tiny amounts of fluoride in the tissue fluids for long periods interferes with the proper growth, development and function of many parts of the body" (Exner et al. 1957, 37).

These conflicts based on different premises appear to have resulted partly from poor communication between adversaries, or from a strong motivation to win the argument. Qualified referees could probably have eliminated them during a public debate so that the technical issues stood out more clearly. There is another source of confusion, however, that is intrinsic to such disagreements. Even with perfect communication among temperate adversaries who eschew rhetorical devices intended simply to put the opposing argument in an unfavorable light, experts may disagree on ambiguous observations and assumptions that cannot be resolved by available objective means.

Ambiguities in Data and Interpretation

The theories, models, procedures, and formulas of science and technology are often believed to allow one trained in their use to calculate an unambiguously correct answer. However, complex technical problems pushing the state-of-the-art often require subtle assumptions that are not easy to articulate. When it is necessary to make a simplifying assumption, and many are reasonable, which one should be made? When data are lacking on a question, how far may one reasonably extrapolate from other data of other sources? How trustworthy is a set of empirical observations? These questions all require judgments, and it is here that experts frequently disagree. I call these points of disagreement "ambiguities."

At the time of the Gofman–Bond debate, most experts agreed that radiation increased the incidence of leukemia and thyroid cancer in an exposed population. However, there was disagreement on whether other forms of cancer are similarly induced by radiation because trustworthy data on these cancers in exposed populations were not available. Consider the plight of researchers trying to calculate risk, or the number of cancers expected in a population exposed to a given level of radiation. Do they calculate an increase in leukemia and thyroid cancer only, or do they calculate an increase in all forms of cancer? Because the former constitute about 10% of all cancers in the United States, the two calculations will differ by about a factor of 10.

And indeed, Gofman's and Bond's calculations of risk values (Table 4-1) differ by a factor of almost 10 (0.94 versus 0.1). The difference arises because Gofman assumed that all radiation increases the incidence of all cancers, whereas Bond assumed that radiation affects only leukemia and thyroid cancer. Rather than acknowledging the limited state of knowledge, the adversaries took a less equivocal position. "Dr. Gofman's excessive estimates are based on the untenable assumptions that all forms of cancer are increased by exposure at low doses and rates.... These assumptions do not square with the facts," claimed Bond (1970, 2). And on the other side: "[Almost] all the major forms of human cancer were by [1969]...already known to be produced by ionizing radiation.... So it became possible to state a primary principle, or 'law,' of radiation production of cancer in humans. That principle or law states, 'All forms of human cancer are, in all probability, induced by ionizing radiation'" (Tamplin and Gofman 1970, 13).

Each scientist chose to accept as a firm conclusion what others regarded as only a tentative hypothesis. But their conclusions cannot be considered "wrong" in the sense that an arithmetic solution can be wrong. Scientific "truths" are never proved but only gain increasing acceptance (and even then are sometimes found to be incorrect). The point at which a hypothesis becomes a conclusion differs from one scientist to another.

Given the inconclusive nature of the data, it is possible to postulate several different relationships between the radiation dose delivered to a population and the resultant increase in cancer. Presumably, with more complete data, some of these relationships might be shown not to exist.

There are two common models explaining the relationship between the cumulative dose of radiation (from birth) received by a population and the incidence of leukemia in that population. These are the linear and threshold models (Figure 4-1). The first model, favored by Gofman and Tamplin and many other opponents of nuclear power, assumes a simple linear relationship between the dose of radiation received and the incidence of leukemia. Linearity implies there will be some incidence of leukemia no matter how low the dose; the only "safe" exposure is zero exposure. The underlying assumption is that a single x ray is sufficient to disrupt the genetic material in a cell and trigger a cancer. The higher the dose of radiation, the more cells will be disrupted.

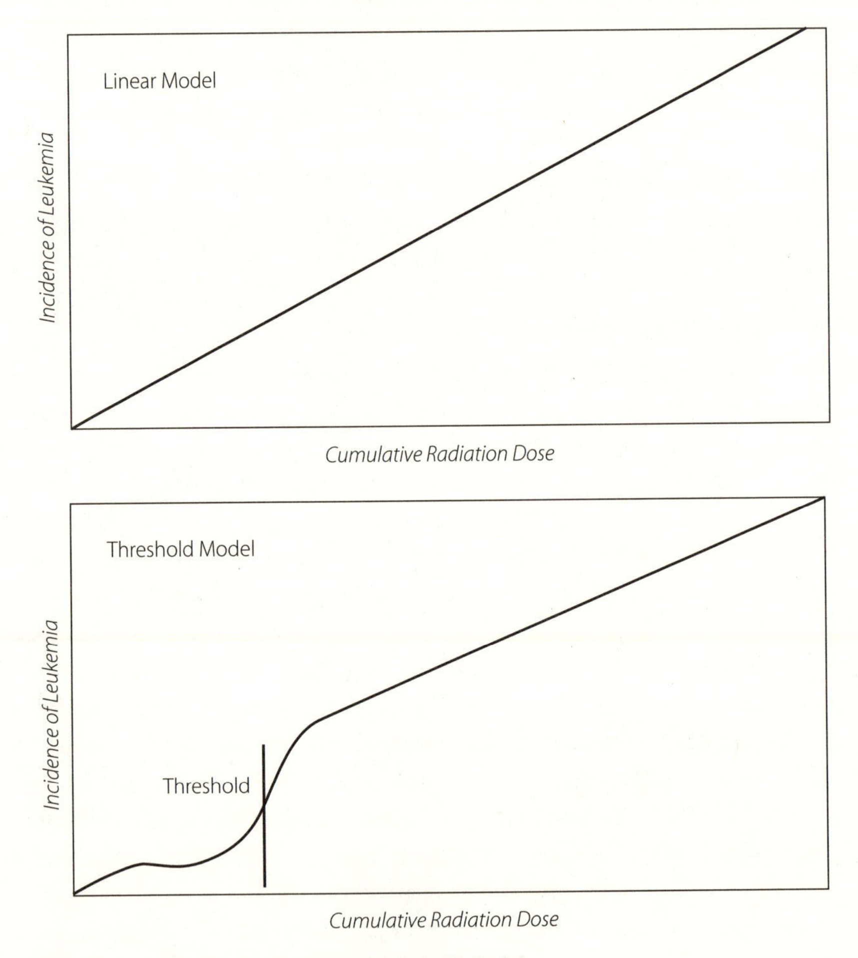

Figure 4-1. Comparison of Linear and Threshold Models

The second model assumes there is a "threshold" dose level below which there is practically no additional incidence of leukemia. The underlying conjecture is that cells have repair mechanisms that prevent occasional breaks in the DNA from disrupting cell functions. Cells become cancerous only when exposure is sufficiently intense to overwhelm these repair capabilities—the threshold level. The implication is that population exposures are safe as long as doses are below the threshold level.

The scanty data on very low exposures are not inconsistent with either model, so it is impossible to say which (if either) is true. This theoretical ambiguity has major implications for the technical debate over permissible radiation standards. One model says there is a safe level of radiation exposure for the population, the other model says there is no safe level. The ambiguous nature of these dose–effect curves is well recognized by radiation biologists. Today the linear model is generally adopted for regulatory guidelines, not because it is true but because it is more conservative for the purpose of public safety.

It would be reasonable, given these ambiguities, for opposing experts to "agree to disagree" and to suspend the debate, at least until new data permit the issue to be reopened. That does not usually occur. Instead, as debate continues, opponents aim their cases not directly at the expert adversary but frequently at a third party such as the public, a congressional committee, or the press. Investing increasing intellect and emotion in the ensuing dispute, adversaries often solidify their cases by using one or both of two polemic strategies that can be applied to ambiguities: rejecting discrepant data and presenting alternative interpretations.

Rejecting Discrepant Data

A common way to deal with data that are inconsistent with one's own position is to deny their scientific validity. In one of many examples from the fluoridation controversy, Dr. Alfred Taylor showed experimentally that when a strain of mice normally susceptible to mammary cancer was regularly given fluoridated water, tumors appeared earlier than in control mice given nonfluoridated water (Mazur 1981). Taylor reported his findings to the Public Health Service, which sent Dr. Howard Andervont to visit Taylor's laboratory. Andervont later testified that Taylor's experiments were not valid. "We came to the conclusion that inasmuch as the food [that Taylor] was feeding to his mice contained 30 to 40 parts per million of fluoride, that the 1/2 part per million [fluorine] in the drinking water could not conceivably have had much influence on his results" (Andervont 1952, 1666).

One could discredit Andervont's rejection of Taylor's analysis by pointing out that fluoride consumed in water is almost completely absorbed into the blood, whereas fluoride consumed as a solid must first be digested, and smaller amounts will reach the bloodstream. Exner and colleagues (1957) made that objection and emphasized that both the experimental and control groups were given the high-fluoride food but that only the experimental group received fluoridated water.

Armstrong and colleagues (1954) conducted an experiment to check Taylor's result. The mean age at which tumors appeared in their experimental mice (which had been given fluoridated water) was lower than that of their control mice. However, since the difference was not statistically significant at the conventional .05 level, they considered the result attributable to chance. Taylor tried to deny the validity of these findings by arguing that Armstrong and his colleagues did not use a large enough number of mice. "A control group consisting of 31 animals would be insufficient to reveal differences of the order of those encountered in the work here" (quoted in Exner et al. 1957, 185). But the control groups in Taylor's own experiments contained an even smaller number of mice. The subjective and selective nature of these attacks and rebuttals is clear.

Presenting Alternative Explanations

Even if both disputants in a technical argument accept the validity of a datum, they may disagree about its interpretation or implication. For example, during initial excavations for a contested nuclear power plant at Bodega Bay, California, an earthquake fault was discovered running through the shaft. It was determined that there had been no movement along the fault for about 40,000 years. Proponents of the plant claimed that the fault was inactive and there was little likelihood of future movement. Opponents claimed that since there had been no movement for a long time, an earthquake was due (Novick 1969).

Statistical data are particularly amenable to alternative interpretations, especially if they contain substantial error variance, as is usually the case in epidemiological studies. Aggregating and disaggregating data can also produce different results. H. Trendley Dean (1938) analyzed the incidence of dental cavities in children in two sets of cities in Colorado and Illinois, one set with water high in naturally occurring fluoride, the other set with water low in naturally occurring fluoride. He concluded that children using water with high fluoride content were freer of cavities than those using water with a lower fluoride content (Table 4-2). Exner, an opponent of fluoridation, analyzed the same data in disaggregated form, noting, "It would appear to take some ingenuity and a certain amount of determination to deduce from these data the conclusion Dean drew" (Exner et al. 1957, 114). Frank J. McClure (1970), a proponent of fluoridation, later used the same data but returned to the aggregated form.

Adversaries in the nuclear controversy treated data on radiation-induced cancer in a similar way to support their positions. Evans (1966) collected data on radium workers showing that no cancers occurred below a median dose level of 0.55 microcuries[2] (Table 4-3). Evans considered this support for the threshold dose–effect curve.

Gofman and Tamplin believed that the same data fit their linear dose–effect curve. Here is their reasoning. First, they estimated the proba-

Table 4-2. Relationship between Fluoride and Incidence of Cavities in Permanent Teeth

Cities	*Number of children examined*	*Range of fluoride (parts per million)*	*Percentage free of cavities*
Dean display			
Pueblo, CO; Junction City, IL; East Moline, IL	114	0.6–1.5	26
Monmouth, IL; Galesburg, CO; Colorado Springs, CO	122	1.7–2.5	49
Exner display			
Pueblo	49	0.6	37
Junction City	30	0.7	26
East Moline	35	1.5	11
Monmouth	29	1.7	55
Galesburg	39	1.8	56
Colorado Springs	54	2.5	41

Source: Mazur 1981.

bility of finding cancer in a person exposed to a given dose of radiation. Focusing on the 5.5-microcuries median-dose group (which has the largest number of cancers and hence is most statistically reliable), they noted 14 cancers out of a total of 40 cases, so there is a 14/40 probability of cancer per person at a median dose of 5.5 microcuries. The probability of cancer per person per microcurie is then $14/(40 \times 5.5) = 0.064$. There are 80 cases with a median dose of 0.055 microcuries, so, assuming the linear hypothesis, the expected number of cancers in that group is $0.064 \times 0.055 \times 80 = 0.28$ cases. However, because cancers cannot occur in fractions, the most likely outcome is zero cancers in this group, which matches what was found. A similar analysis for the lower median doses shows that in each group the expected number of cancers is near zero. Gofman and Tamplin thus argued that the data are fully consistent with the linear dose–effect curve, and that the apparent threshold is due to the fact that very small numbers of persons are exposed.

Alternative modes of interpretations are often used to explain away an opposing argument. Finding no fluoride poisoning in cities with naturally fluoridated water (at about one part per million), supporters of fluoridation concluded that low concentrations of fluoride must be safe. Critics of fluoridation argued that low concentrations of fluoride would cause poisoning, but noted that calcium is an antidote. Since calcium usually occurs naturally in the same waters where fluoride occurs naturally, this explains the lack of poisoning (Exner et al. 1957).

Table 4-3. Relationship between Exposure to Radiation and Incidence of Cancer in Radium Workers

Median dose (microcuries of radium equivalent residual)	*Number of cases*	*Number of cancers*
<0.001	42	0
0.0055	61	0
0.055	80	0
0.55	32	3
5.5	40	14
55.0	14	2

Source: Mazur 1981.

Proponents of nuclear power minimized the harmful effects of long-term, low-level radiation based on animal experiments indicating that a given dose of radiation over a protracted period is less harmful than the same dose delivered in a short period of time. In a typical experiment of this sort, one group of 10-week-old mice receives small doses of radiation daily. A second group of 10-week-old mice is given the same total amount of radiation in a single dose. The incidence of cancer is usually higher in the second group.

Tamplin and Gofman dismissed the mitigating effects of protracted dosage by pointing out that for a given dose and dose rate of radiation, more harm would occur in a younger organism than in an older one. The mice receiving acute doses were fully irradiated at 10 weeks of age, whereas the protracted group received most of its radiation after 10 weeks. Therefore, the lower incidence of cancer in the protracted group simply reflected the fact that the mice were older when irradiated.

RECOGNIZING STRUCTURAL BIASES

Scientific claims made by experts in controversies are fairly predictable from their policy positions. Opponents of nuclear power usually favor the linear model of radiation effects; proponents more likely argue for a threshold model. Experts opposing fluoridation claim that fluoride is toxic and that any exposure is potentially harmful. Experts favoring fluoridation claim that the toxicity of any substance, whether fluoride or table salt, depends on how much is consumed—the dose makes the poison—and that one part per million of fluoride is too low a concentration to produce toxic effects. In general, proponents of a technology minimize its risks and maximize its benefits; opponents do the opposite.

Less obvious but equally important is the existence of what sociologists call "structural biases." These are biases, whether political or technical, induced by an expert's position in the profession's social structure: his or her field or discipline, place of employment, academic prestige, and gender. None of these structural variables has any direct relevance to the scientific or policy issues at hand. Nonetheless, they do influence scientists' views, as illustrated by a large survey of political attitudes among

American professors conducted during the 1960s (Ladd and Lipset 1976; see Hamilton and Hargens 1993 for a partial but more recent replication). The survey asked respondents their opinions on various contemporary social issues. Based on their responses, they were categorized as politically liberal, moderate, or conservative. Table 4-4 shows the percentage of "liberal" professors, broken down by academic discipline and quality of university (measured in quartiles, from "elite" to "lowest"). There is considerable variation in political orientation across disciplines, with the average physics professor more liberal than the average professor of engineering. (No technical discipline was as left leaning as the social sciences, where 64% of professors were "liberal.") Across disciplines, professors at higher-status universities were more liberal than those at lower-status universities.

During the 1960s and 1970s it was common for scientists and professors to circulate petitions favoring various political positions and to publish them in the *New York Times*. A large petition opposing nuclear power plants appeared in 1975. The following year a smaller group of elite scientists responded with a petition in support of the technology (Mazur 1981).

Table 4-4. Percentage of "Liberal" Professors, by Discipline and Quality of University

		Quartile			
Discipline	*Total*	*Elite-level university* 1	2	3	*Lowest-level university* 4
Physics	45	66	47	28	33
Medicine	38	42	30	26	NA
Mathematics	36	67	45	29	18
Biology	35	54	36	26	24
Chemistry	35	52	42	26	28
Geology	32	46	32	25	17
Engineering	24	29	20	17	16

Source: Adapted from Ladd and Lipset 1976.

These dueling petitions are good grist for the sociological data mill, providing a glimpse of the structural underpinnings of the nuclear power debate. Biomedical scientists were particularly likely to oppose nuclear power plants. Members of the prestigious National Academy of Sciences were more likely to support the plants, as were scientists who had signed earlier *Times* petitions favoring the presidential candidacy of Lyndon Johnson or the deployment of an ABM system (Mazur 1981). The combination of these factors works well in differentiating scientists who signed pro- and antinuclear petitions (Figure 4-2). Similar factors predict positions in the debate over fallout from nuclear weapons testing (Kopp 1979).

If structural biases affect political beliefs, do they affect scientific beliefs too? Carcinogenesis offers a good test case, because the extent to which synthetic substances contribute to human cancer has not been definitively established. Lichter and Rothman (1999) conducted telephone

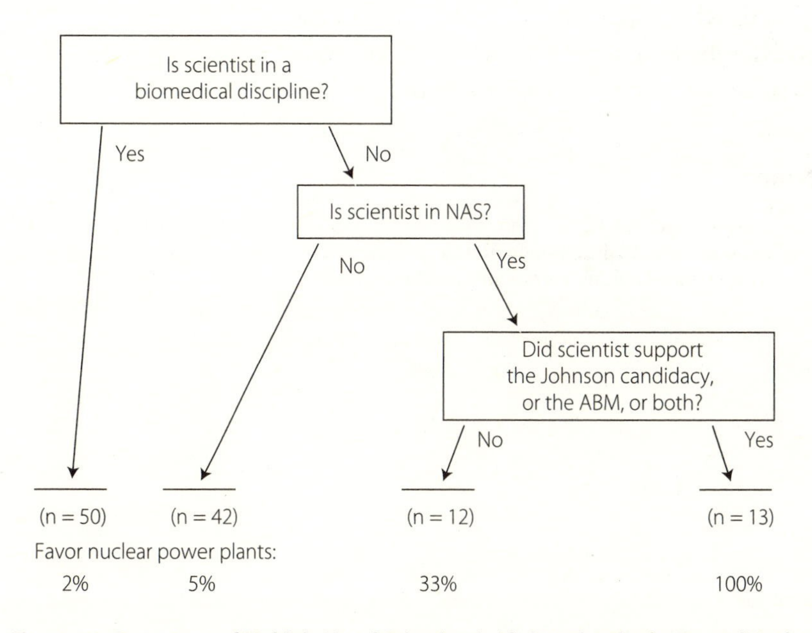

Figure 4-2. Percentage of Multiply Listed Scientists (with Stated Attitudes Toward Nuclear Power) Who Favor Nuclear Plants, as a Function of Three Social Characteristics

interviews with 401 cancer researchers selected randomly from members of the American Association for Cancer Research listing their specialization as carcinogenesis or epidemiology. Each interview posed 90 questions about potential contributors to U.S. cancer rates and scientific controversies related to cancer research.

Professor Stanley Rothman has kindly given me access to the responses, and my analysis shows systematic differences among researchers on the importance of synthetic causes of cancer, on whether the United States is facing a cancer epidemic, on the hazards of trace doses, and on the adequacy of the mass media in warning about synthetic carcinogens. These views bundle together, with researchers at one pole perceiving the nation under dire threat from industrial activity and synthetic chemicals, and those at the other seeing synthetic causes as unimportant compared with smoking and natural carcinogens. Cancer researchers who are male, who are affiliated with a medical school, and who have many publications are more likely than their colleagues to minimize the importance of synthetic carcinogens (Mazur et al. 2001).

Subjective judgments of scientists and engineers are shaped by their social milieus; they are no different in this regard from anyone else. How, then, do we know when to believe them?

Notes

1. Rem (roentgen equivalent man) measures the biological damage of radiation. It takes into account both the amount, or dose, of radiation and the biological effect of the type of radiation. A millirem is one-thousandth of a rem.
2. A curie is a unit of radioactivity. A curie is the amount of a radionuclide that undergoes 37 billion atomic transformations per second. A microcurie is one-millionth of a curie.

CHAPTER 5

EVALUATING THE LAWLESS WARNINGS: TRUE OR FALSE?

In this chapter, I evaluate whether each of Lawless's cases proved true or false in light of 30 years of further knowledge. Given the caveats I presented in the previous chapter, this task may seem intellectually untenable. Nearly all public alarms are controversial, and opponents of a technology predictably appraise risks as higher (and benefits lower) than proponents do. Feuding experts have ample room to disagree legitimately over the existence or severity of a hazard. Making matters worse, ideological differences between experts, while seemingly unrelated to the scientific question, may encourage conflicting claims. One socially constructed claim is as valid as its opposite, critics of my approach might argue, so any choice between them comes down to politics.

These critics would be correct in an absolutist sense. To completely discount a risk, one would have to show that there is no probability, however small, of human harm, however slight. It is impossible in principle to unequivocally demonstrate the absence of an effect, even with high-quality data. In practice the problem is even worse, because real-world risk assessments are constrained by limits on time and money. They cannot approach the validity that could be achieved if resources for research were unlimited.

If we cannot obtain objective certainty, can we at least demand consensus among experts that a warning is true or false? The answer is no, because we often find scientific holdouts who disagree with nearly all other experts in evaluating a warning. Recalcitrants need not be poor scientists. Albert Einstein famously opposed quantum theory's interpretation of reality, denying that "God plays dice with the universe" until the day he died, perhaps then learning the truth. Some competent scientists today deny that HIV causes AIDS, that the Earth is more than one billion years old, that humans evolved from other species, that "cold fusion" does not work, and that perpetual motion machines are impossible. In rare cases, these holdouts surrender their positions or are eventually acknowledged to have been correct.

Despite these challenges, I rely on the scientific and regulatory criteria I describe below to determine whether Lawless's sample of warnings proved true or false. Readers can judge for themselves whether my approach and judgments seem valid.

THE CONTINUING DEBATE OVER FLUORIDATION

Most of Lawless's warnings are no longer actively contested, their adversaries having died or lost interest. However, since the conclusions of past debate are usually pregnable, argument is occasionally rejuvenated, rather like old-time political leftists and rightists reawakening the Alger Hiss espionage case from the Cold War. Clever debaters can argue for or against any given warning and then, if they lack conviction, switch sides and defend the opposite stance. Counterarguments—substantive or rhetorical—can usually raise doubt about any conclusion. Such claims deflect attention from cogent scientific arguments. But even when we restrict our focus to rational evaluation of evidence, sometimes there seems to be no progress toward consensus.

Fluoridation, the longest-running technical controversy in the public eye, is the most stubborn of Lawless's selected warnings in reaching scientific closure. Debate continues today, less visibly than in the 1950s and 1960s but no less adamantly. Like nearly all public technical disputes, this one is partly about risk to health and partly about political values, particularly the tension between governmental authority and individual freedom of action. Perhaps because of its association during the McCarthy era with fears of domestic communism, the antifluoridationist position has attracted its share of extreme ideologues and continues to do so. National Public Radio recently reported on a fluoridation decision in Arizona featuring one opponent who insisted that Hitler had added fluoride to Germany's water to make the population docile (Morning Edition, April 27, 2001). A plausible debate between two fictitious scientists illustrates the arguments on both sides of this controversy.

Profluoride: Authoritative scientific and medical organizations have repeatedly stated that fluoridation at the recommended level of 1 part per million in water is safe. In 1993, a committee of the National Academy of Sciences reviewed the health effects of ingested fluoride for the Environmental Protection Agency (EPA). The committee found no convincing evidence that exposures below 4 parts per million in drinking water produce kidney disease, gastrointestinal or immune system problems, adverse reproductive effects, genotoxicity, or cancer. Fluoride at recommended levels does produce some dental fluorosis, usually as a barely discernible white spotting on the enamel. The effect of fluoride on bone strength and hip fractures has been addressed in experimental studies on humans and animals, yielding inconsistent results, some showing a weak association between fluoride in drinking water and the risk of hip fracture. The committee found no basis for recommending that EPA lower the current standard for fluoride.

Antifluoride: That National Academy of Sciences study discounted research suggesting that fluoride is harmful at recommended levels. For example, it notes studies showing the risk of hip fracture increases with the level of fluoride in drinking water but then ignores the implication. Low levels of fluoride may alter bones' crystalline structure and density, changes that are harder to detect than fractures but that may eventually weaken the bone or impair its function. No one doubts that fluoride is toxic at high doses or that it causes at least one toxic effect—dental fluorosis—at the recommended level of 1 part per million. Fluoride may be especially toxic to extremely sensitive people. For example, diabetics who drink a lot of water and people on dialysis may either consume more fluoride or be more vulnerable to its toxic effects. Possible adverse effects on these sensitive people have not been fully evaluated.

Profluoride: Epidemiological studies are inexact and have many sources of error, so two studies can give differing results. These studies give no consistent picture of bone fractures increasing with fluoride concentration. Mild enamel fluorosis is a cosmetic effect, not a toxic effect. Allergists and immunologists have not accepted the existence of "fluoride-sensitive" people. It is impossible to evaluate every conceivable consequence in every subgroup of the population.

Antifluoride: How can anyone disprove the existence of fluoride-sensitive people? To call dental fluorosis a cosmetic effect rather than a toxic effect is playing with words; any change in the body is a potentially adverse change. If we cannot rule out all possibly ill effects, we should not impose lifelong fluoride exposure on healthy people.

Clearly, no resolution is in sight. Anyone requiring full consensus or absolute certainty must remain agnostic about fluoridation, and indeed about virtually all other warnings.

TOXIC CRANBERRIES: HOW IS RISK ASSESSED?

The warning about tainted cranberries is the opposite of fluoridation, in that no one today defends it. Indeed, it is doubtful that Flemming and his scientific advisers thought that the cranberries posed a real health threat. Flemming claimed at his press conference, and his wife reiterates in her memoirs (Flemming 1991, 202), that the recently enacted Delaney Clause, interpreted as banning from food any chemical causing cancer in animals, dictated the warning. (In 1996 legislation removed pesticide residues from the purview of Delaney.) As if in apology to growers, the *New York Times* of November 27, 1959, carried a front-page photo of Mrs. Flemming offering cranberries to her husband at Thanksgiving dinner.

Those who denied the cranberry threat made a qualitative judgment: traces of amitrole were found on a small portion of the crop, but consumption was limited to holiday meals. There seemed intuitively to be little risk.

Modern risk assessments are more quantitative (Kokoski et al. 1990; National Research Council 1994). For all types of toxic effects other than cancer, regulatory agencies generally assume there is a "safe" amount of a chemical to which a person can be exposed on a daily basis over a lifetime without suffering deleterious effect. This level of exposure is usually called the acceptable daily intake (ADI), although terminology varies across agencies. The first step in estimating an ADI is to find the no-observed-effect level (NOEL) in experimental animals (when human data are lacking). This is the exposure level at which there is no statistically significant increase in the frequency or severity of any effect in the exposed population relative to controls. The NOEL approximates the threshold for safe exposure for these animals, below which toxic effects are unlikely. To calculate the ADI for humans, the NOEL is divided by a safety factor—usually 10, 100, or 1,000—intended to provide a cushion for species differences (humans are assumed to be more sensitive than test animals), and for people who may be especially sensitive to the chemical. Regulatory agencies consider daily exposure below the ADI to be acceptable.

For cancer, it is assumed that no threshold exists. Risk estimation for carcinogens therefore follows a different procedure from that for noncarcinogens. First, the relationship between cancer incidence and exposure is determined, usually by studying animals exposed to high doses. Then this relationship is extrapolated downward (usually with a straight line through the origin) to the lowest level of human exposure in order to estimate the lifetime cancer risk from the agent. By assumption, the only "safe" dose is a zero dose. However, if the calculated risk of an excess cancer is very low, usually 10^{-6} or less, it may be considered negligible.

Research reports since 1959 have questioned the significance of thyroid tumors produced in animals by amitrole, either because very high doses were used or because of doubts that the tumors were cancerous. The chemical seems more likely to produce goiters than cancers in the thyroid (Paynter et al. 1988; Hurley 1998). The World Health Organization downplays the cancer risk, but it calculates an ADI for amitrole using a high safety factor of 1,000 because the herbicide does induce thyroid growths, though "presumably via a secondary tumorigenesis mechanism" (Lu 1995). This ADI is 0.0005 milligram per kilogram of bodyweight per day, which implies that a 10-kilogram (22-pound) child can safely consume 0.005 milligram of amitrole per day.

During the cranberry crisis, the FDA considered berries with more than 0.15 part per million amitrole contaminated. If we assume that tainted berries contained three times that level, or 0.45 parts per million, then a child weighing 10 kilograms would have had to ingest 0.01 kilograms of cranberries to reach the acceptable limit. This is less than half an ounce, an amount easily exceeded by a small child at Thanksgiving dinner. But in 1959, before cranberry juice was popular, consumption of the fruit was virtually limited to the holiday; the average exposure over several days would have been trivial. This quantitative result is thus consistent with intuitive impressions that Secretary Flemming's warning was unjustified.[1]

But there is another scientifically legitimate point of view, the "single hit" hypothesis, which claims that one molecule of a carcinogenic chemical (or one x ray) is sufficient to disrupt the genetic material in a cell and initiate a cancer (see Chapter 4). No one doubts that some cranberries contained amitrole, which the World Health Organization's authoritative International Agency for Research on Cancer calls "possibly carcinogenic" in humans (1987). With 100 million people consuming cranberries on Thanksgiving 1959, who can say with confidence that 10 or 100 or 500 celebrants would not have had a cancer triggered that day? No one has made this argument to me because today's experts have no interest in justifying Secretary Flemming's cranberry fiasco.

DECIDING TRUE OR FALSE? RATING THE CASES

Because absolutely certain assessments of risk are impossible, I accept lower standards. For this study, I rate each warning as "true" or "false" using the most recent evaluation by orthodox sources of threats to human health. I depend on conventionally authoritative voices, such as the research committees of the National Academy of Sciences and its operating arm, the National Research Council. I am not oblivious to dissident voices but generally do not substitute them for more orthodox evaluations. (On this basis, I accept HIV as the cause of AIDS despite a few prominent scientists who reject that explanation.)

I use two types of criteria for an authoritative and orthodox evaluation: empirical and regulatory. I give priority to empirical evidence when it shows with reasonable certainty that a threat is valid or false. If possible,

I rely on standard summaries of empirical health effects from exposure to various chemicals, including the EPA's Integrated Risk Information System (http://www.epa.gov/iris/) and the monograph series of the International Agency for Research on Cancer.

Efforts to empirically evaluate alleged hazards in Lawless's sample vary greatly. For example, there are enormous scientific literatures on fluoridation and ionizing radiation, whereas minimal resources have been devoted to risk assessments of enzyme detergents or DMSO. These differences reflect both political judgments (the extent to which people care about a warning) and scientific judgments (the extent to which concrete evidence suggests cause for concern).

The first warning about the "wonder drug" DMSO, coming as its uncontrolled use was rapidly increasing, was not based on any specific evidence of harm but on a cautionary analogy to the thalidomide tragedy three years earlier: perhaps, argued critics, an adverse effect will be discovered after exposure is widespread. Doubts about DMSO were strengthened by evidence of ocular changes in laboratory animals exposed to sustained high doses, but this effect was not replicated in clinical trials on human subjects. In the end, no extensive body of research was ever produced "proving" DMSO's safety, but neither was there any sustained reason to think it harmful in clinical application. Today DMSO seems generally regarded as a harmless substance. Enzyme detergents are another case in which the initial warning of an effect on human health lacked sufficient substance to elicit a voluminous assessment literature.

For some alleged hazards, such as exposure to low-level x rays, empirical evaluation is extensive but nonetheless remains indeterminate. In such cases, if there is accepted regulatory wisdom that exposures are or are not to be treated as harmful, I accept that as an inferred evaluation of the warning. For example, it is widely accepted today that for regulatory purposes, ionizing radiation should be assumed to have deleterious effects even at the lowest doses and should therefore be avoided unless there are compensating benefits. By this regulatory—not empirical—criterion, low-level x rays are rated as a true threat (see Goldman 1996 for an opposing position).

Sometimes safe substances are regulated while unsafe substances are not. Regulatory decisions may depend as much on politics or cost–benefit considerations as on a validated hazard to human life. Therefore, regulation itself is no proof of risk, nor can one equate failure to regulate with the absence of risk. Still, I think it reasonable to take the presence or absence of regulation as a pragmatic decision by society to treat a warning as true or false. Empirical and regulatory criteria are usually consistent, as in the case of fluoridation, which both approaches indicate is a false warning.

My evaluations are limited to warnings about human health and do not include threats to animals or the environment. A false warning about humans may be true regarding wildlife or environmental degradation. I consider as a separate issue whether or not a warning was wise or justified at the time it was raised, regardless of how the risk is evaluated today. One might reasonably hold that it was wise in 1950 to warn against fluoridation even though U.S. authorities subsequently accepted fluoridation as a safe procedure (see Appendix II).

Table 5-1 summarizes my assessments of the 31 warnings as true or false based on empirical and regulatory criteria. Case summaries in Appendix I provide supporting details. I do not use both criteria for every case. Sometimes regulatory standards are absent or immaterial. For example, the warning by health officials that contaminated cans of soup could cause botulism was validated empirically by the discovery of additional cans of contaminated soup. Although federal law prohibits botulin in canned food, this long-standing regulation was irrelevant to the crisis at hand, so I ignore it.

Some warnings are difficult to test, such as those about very low exposures to chemicals or radiation and may never be settled empirically. Other warnings could be tested adequately if more resources were devoted to the task. To avoid extended debates over ambiguous empirical results, I apply a variant of Occam's razor, ignoring the empirical criterion when regulation decrees that a warning is to be treated as true or false. The right-hand column of Table 5-1 shows my final assessment of each warning.

Cyclamate is the single case in which the empirical and the regulatory criteria differ. In 1970, citing the Delaney Clause, the FDA banned cyclamate because one study of rats fed a combination of cyclamate and saccharin showed an increased rate of bladder cancer. Two attempts to replicate the study failed, and a considerable body of research has since

Table 5-1. Assessment of Warnings, Based on Empirical or Regulatory Criteria

Case		*Empirical criterion*	*Regulatory criterion*	*Final assessment*
1.	Oral contraceptives	True	–	*True*
2.	Contaminated cranberries	False	–	*False*
3.	DES in livestock	–	True	*True*
4.	Cyclamate	False	True	*False**
5.	MSG	False	False	*False*
6.	Botulism	True	–	*True*
7.	Fish protein concentrate	–	False	*False*
8.	Fluoridation	False	False	*False*
9.	Salk polio vaccine	True	–	*True*
10.	Thalidomide	True	–	*True*
11.	Hexachlorophene	–	True	*True*
12.	DMSO	False	–	*False*
13.	Shoe fluoroscopes	–	True	*True*
14.	Medical x rays	–	True	*True*
15.	Radiation from defective televisions	–	True	*True*
16.	Smog in Donora	True	True	*True*
17.	Mercury pollution from industry	–	True	*True*
18.	Mercury in tuna	–	False	*False*
19.	DDT	–	True	*True*
20.	Asbestos	True	True	*True*
21.	Taconite pollution	–	False	*False*
22.	Enzyme detergents	False	False	*False*
23.	NTA in detergents	False	False	*False*
24.	Plutonium at Rocky Flats	True	True	*True*
25.	Radioactive waste stored in Kansas	True	–	*True*
26.	Nuclear test on Amchitka	False	False	*False*
27.	Poison gas released at Dugway	True	–	*True*
28.	Nerve gas disposal	–	True	*True*
29.	ELF radiation at Project Sanguine	False	–	*False*
30.	Chemical Mace	False	False	*False*
31.	Injuries on synthetic turf	True	–	*True*

Note: A dash (–) means "not applicable."

* Criteria are inconsistent.

shown that cyclamate does not cause cancer or birth defects in mammals, as once feared. A petition by manufacturers to reapprove the additive was submitted to the FDA in 1989 but is still pending, and the agency cannot comment on pending matters. More than 50 nations, including most of Europe, allow cyclamate as a sweetener. The World Health Organization's Joint Expert Committee on Food Additives has consistently determined cyclamate to be safe. Canada, in a strange twist from the U.S. position, has more stringent restrictions on saccharin than on cyclamate, allowing both as tabletop sweeteners but only cyclamate as a sweetening agent in drugs (Appendix II). Contrary to the U.S. regulatory position, I accept the empirical evidence that the warning about cyclamate was a false alarm.

I focus on taconite rather than asbestos in assessing the warning against Reserve Mining Company's pollution of Lake Superior (Case 21) because Lawless mentions asbestos only peripherally, entitling this case "Taconite Pollution of Lake Superior." (The ensuing lawsuit is a classic case in environmental law.) In 1969 the first public warning was raised over dumping of taconite (iron ore) tailings into the lake, threatening the safety of drinking water drawn by cities along the coastline. Without doubt taconite tailings polluted the lake, but there is neither empirical nor regulatory support for the claim that taconite in drinking water poses a health threat.

The problem took on a new dimension in 1973 when the EPA announced that Duluth's drinking water contained asbestoslike fibers in amounts of up to 100 billion fibers per liter. (EPA currently limits asbestos in drinking water to seven million fibers per liter for fibers larger than 10 micrometers.) I have limited this warning to taconite, regarding it as a false alarm. In the next chapter I describe how I will further code Lawless's cases to analyze them for hallmarks of true and false warnings.

Note

1. A similar calculation can be made for fluoridation. The EPA's "reference dose," essentially the same as an ADI, for soluble fluoride is 0.06 milligram per kilogram of body weight per day, based on epidemiological studies of children. In children consuming more than that amount, mottled dental enamel is observed, which the EPA regards as a cosmetic effect. With water fluoridated at the recommended 1 part per million, a child weighing 20 kilograms (44 pounds) reaches the ADI by drinking 1 liter of water and eat-

ing 0.01 milligrams of fluoride in food (www.epa.gov/ngispgm3/iris/). Thus, the "acceptable intake" of fluoride is much closer to actual consumption than was the "acceptable intake" of contaminated cranberries.

CHAPTER 6

CODING THE CASES

The Lawless project team initially identified more than 200 episodes since World War II in which technological information reported in the popular news media had concerned or alarmed segments of the public. The subsample of 45 cases Lawless finally chose for detailed study was intended to be representative, but no specific criteria for inclusion were closely followed. Excluded were routine problems such as transportation accidents and industrial fires, occupational hazards such as mine disasters and other unsafe working conditions, and problems involving government security (except the Dugway sheep kill).

My own sample contains the 31 of Lawless's detailed cases that involve public warnings about technological threats to human health (and sometimes additionally to animals and the environment). Of the 14 Lawless cases I have excluded from further analysis, 9 concern environmental threats only, and 5 are about ethics or fraud.

Obtaining a representative sample of technological issues that have evoked public alarm is difficult. There is no well-defined universe of technological problems from which a random sample can be drawn. My impression of Lawless' sample is that it is diverse but weighted toward issues involving the FDA while excluding some of the most highly publicized technological warnings of the era, particularly those concerning radioactive fallout and nuclear weapons, nuclear reactors, antiballistic missiles, and the supersonic transport. Nonetheless, his project team reached a reasonable solution to a problem that cannot be definitively solved. Most importantly, the Lawless team had no inkling of my hypotheses, so whatever biases may have affected its selection of cases, we may be sure they do not purposively (or inadvertently) favor results I report in Chapter 7. Had I drawn the sample myself, even with the highest aspiration for objectivity, the results would not be credible.

In the remainder of this chapter I define the variables I used to code the cases to enable me to analyze them for hallmarks of true and false warnings. With the help of research assistants, I reexamined each of the 31 warnings, reading contemporaneous press accounts and other reports, examining recent documentation, and conducting telephone or email interviews. We accumulated considerable—sometimes voluminous—material for each case.

When variables used in a content analysis such as this one can be coded with minimal effort, two or more independent coders should rate them to ensure reliability. However, in this study, a professionally competent person required days to independently digest and code a typical case. Thus, given limits on time and money, I depended instead on two other means of quality control. Where practical I used Lawless's coding and his other judgments rather than my own, knowing that whatever his prejudices, they were independent of my hypotheses. If a code seemed problematic, I sought critical reviews from multiple experts before finalizing the result. Colleagues or other informants have vetted nearly everything in this study.

Summaries of all cases, including codes, appear in Appendix I. Each summary begins with the case number (in order of appearance in Lawless [1977]), a shorthand label, and a brief description of the warning. It also includes the following variables.

- First public warning.
- Type of news source for first public warning (scientific, government, or citizen).

- Did source have an identifiable preexisting bias?
- Risk assessment: Are warnings true or false?
- Lawless's assessment of whether a threat was overemphasized or not.
- Was media coverage hyped or routine?
- Lawless's media score.
- Was warning derived or not?

In the next chapter I use the coded cases to evaluate my hypotheses, displaying my results in a list format (rather than traditional contingency tables) to increase the transparency of my analysis. Readers may easily appraise how my results might have changed given different coding of particular cases.

FIRST PUBLIC WARNING

As a first step in coding Lawless's cases, I determined the source of the first conspicuous public report of a warning. With a few exceptions, I defined such a source as publication in a widely read, nationally distributed newspaper or magazine of an easily noticed—not tiny or buried—report. The "Key Events and Roles" section of Lawless's case description usually includes this information, or it is easily inferred. Where his text does not clearly identify the first warning, I used the earliest news report cited in his bibliography. If that was not adequate, I searched the *New York Times Index* to identify (or verify) early news coverage. I did not count warnings in scientific, medical, environmental, or other specialized media.

I found most "first" warnings in the *New York Times*, partly because it is regarded as the nation's "newspaper of record," but also because the *Times* and its *Index* together constitute the most complete and accessible news archive. No doubt I missed some earlier notices in other news organs. However, this is not a serious problem because early news accounts typically convey a similar message and rely on the same sources of information.

In three cases (fluoridation, the Rocky Flats plutonium plant, and the Dugway sheep kill), the national news media were inexplicably slow to report issues that had provoked considerable public protest and local news coverage. National stories, when they finally appeared, reported warnings already known to the relevant audience. In these cases I identified local news as the "first" public warnings.

Sometimes a public warning changed with time into a considerably different warning from that first raised. In such cases I focused on the aspect that was most prominent during the study period or that made the most substantive sense. For example, the first warning about DES, in 1957, concerned its use in fattening chickens and other livestock, raising fears that DES residues in poultry and meat are hazardous to humans. The subsequent warning about DES as a human drug was not raised until 1971, following the discovery of vaginal cancer in the daughters of women who had taken DES while pregnant to avoid miscarriage. I addressed the first warning about DES, its use to fatten livestock, not its use as a human drug. (The pharmaceutical warning is empirically true; the fattener warning is true on regulatory grounds, given that regulators have prohibited use of DES as a fattener.) The first health warning about DDT, reported in the *New York Times* in 1949, was a physician's claim linking DDT to "virus X" in humans. Because this warning was an anomaly, I ignored it in favor of Rachel Carson's 1962 warning in *Silent Spring* of other health effects from the insecticide.

TYPE OF NEWS SOURCE

In coding the type of news source, my guiding principle was to identify the voice the journalist regards as the primary source of information. Often this is clear from the headline, for example, "Scientists Term Radiation a Peril to Future of Man," which introduced a *New York Times* report on the excessive use of medical x rays (June 16, 1956, 1). I coded this primary source as scientific, government, or citizen-activist.

I coded the source "scientific" if it has the accoutrements of conventional science. These warnings are based on scientific results established by generally approved methods, performed at a conventional scientific or medical institution, and reported in a legitimate scientific outlet. I also indicated whether the research results appear in a peer-reviewed scientific publication at the time of the warning.

This coding of news sources as scientific, government, or citizen is intended to separate activities of a research character from those concerned with public administration or influencing governmental or corporate policy. Warnings originating from a government agency or official, at any level, are coded "government," whether or not scientists were on the staff of the government body. Warnings coming initially from a citizen activist or advocacy group were coded "citizen," whether or not scientists belong to the organization. For example, the warning about a radioactive release from the Rocky Flats plutonium plant was raised by the Colorado Committee for Environmental Information (CCEI), a voluntary organization made up mostly of physicists, who had surveyed radiation levels near the facility. Because CCEI played an advocacy role, participating actively in litigation and influencing public policy, I designated it a "citizen" rather than a "scientific" source.

Coding is not always 100% clear-cut, as illustrated by the case of hexachlorophene (HCP). The *New York Times* headlined the first public warning about HCP "Detergent Solution Used in Baby Baths Viewed as a Peril" (August 22, 1971, 28), basing its report on an article in the peer-reviewed British medical journal the *Lancet*. The article had five authors. Three worked at the FDA's Toxicology Branch and had begun their work on HCP because of a manufacturer's application to use the chemical as a fungicide. The other two authors were nongovernment scientists employed at the Albert Einstein College of Medicine. The *Times* cites the *Lancet* as its source and does not mention the FDA. No government agency had issued any warning at that time. (Indeed, the FDA was later criticized for ignoring HCP [Wade 1971].) Thus even though three of the five authors worked for the FDA, they conducted and published their research in a way indistinguishable from routine activity at a university or medical institution. Therefore I coded the source "scientific, peer reviewed." Readers who prefer a "government" designation may revise Table 7-1 in the next chapter accordingly.

Another ambiguous source is the Kansas Geological Survey (KGS), part of the University of Kansas and funded by state government, which has characteristics of both a scientific and a government body. To resolve its designation, I queried the Survey, receiving the following reply.

> The KGS is a division of the University of Kansas. We are a line-item in the University's budget. We don't have any regulatory responsibility. In that sense, we are strictly a university operation. By the same token, many people in state government, and in the state of Kansas, treat us as a state agency, and in some senses, we certainly fill the role of a state agency. But technically we are strictly a university entity. (Rex Buchanan, Associate Director for Public Outreach, pers. comm.)

Given my intent to distinguish sources with a research character from those essentially concerned with public administration or with influencing governmental or corporate policy, it seems proper to call the Survey a scientific source.

In some cases, it is important to distinguish a proximate news source—the person who talks to a journalist—from a more distant source of information. When the *New York Times* reported that underground salt formations in Kansas might be unsafe as a site for permanent storage of radioactive waste, it used information supplied by Ronald Baxter, chair of the Kansas Sierra Club. Baxter had given the *Times* a critical evaluation of the site's geology, produced by the Kansas Geological Survey. My intention was to identify the source that appears as the authoritative and substantive voice in the news article. Here the primary voice was the Kansas Geological Survey, not Baxter, the citizen-activist who was the conduit to a reporter. This is manifest in the article's headline, "Kansas Geologists Oppose a Nuclear Waste Dump" (February 17, 1971, 27).

The warning about monosodium glutamate (MSG) in baby food, also carried in the *New York Times*, is a contrasting example. Here the proximate source was consumer advocate Ralph Nader, who brought the warning to press attention during testimony in congressional hearings. Nader seems to have based his warning on a 1969 report in the journal *Science* by J. Olney that mice treated with MSG suffered brain lesions. The journalist nevertheless made Nader the authoritative and substantive voice in the *Times* article, as shown in its headline, "Nader Questions Safety of Baby Food Additives" (July 16, 1969: 51). The scientific study reported in *Science* is not mentioned.

The news source for one warning resists classification as either scientific, government, or citizen. It is a rare example of a corporation, General Electric, warning the public about one of its own products, a defective color television emitting an unacceptable level of x radiation. GE had tried to deal with the matter quietly, but upon learning that the story had been leaked and was about to appear in the *New York Times,* it made the first (preemptive) public announcement. (For details, see the summary of Case 15 in Appendix I.) I coded the type of news source as "General Electric," including it or omitting it from data displays as appropriate.

DID THE SOURCE HAVE A PREEXISTING BIAS?

I coded a warning "yes" for bias if I found evidence that the person or group raising the warning showed prior hostility or negativism toward the party blamed for the hazard. Lacking such evidence, I assumed the warning was free of prejudice. Of course, bias may be hidden, in which case my coding is erroneous. But bias is unlikely to remain hidden, because opposing sides in a dispute can usually be counted on to reveal their opponents' prejudices. Charges of bias, if at all justified, become visible in the public debate. Ralph Nader, for example, is perennially identified as anti-corporation to discredit his warnings about specific products.

RISK ASSESSMENT: ARE WARNINGS TRUE OR FALSE?

Based on the information available today, I coded each case as true or false. I also indicated whether the assessment was based on empirical or regulatory grounds. (See Chapter 5.)

DID LAWLESS REGARD THE THREAT AS OVEREMPHASIZED?

Lawless provides succinct assessments of the validity of 27 warnings based on the scientific knowledge available to his project team in the 1970s. These appear in his Appendix A-XVI, where he enters a "yes" or "no" under the label "Threat Was Overemphasized by 'Opponents' of the Technology." A "yes" means the project team considered the threat exaggerated, a "no" means it seemed valid. Unfortunately, Lawless included no explanation of these summary judgments, and he included no coding for MSG, botulism, fluoridation, or DDT.

WAS MEDIA COVERAGE HYPED OR ROUTINE?

I coded media coverage of each warning as "routine" or "hyped." "Routine" coverage occurs when a story receives an ordinary amount of space, duration, and salience. "Hyped" coverage occurs when a source or journalist uses unusual or clearly exaggerated information or extraordinary visuals to heighten news coverage or intensify alarm and sensationalism. The warning against cyclamate is coded "hyped" because it became a public issue when a mid-level scientist from the Food and Drug Administration appeared on television news, shortly after the thalidomide incident, displaying deformed chicken embryos produced when cyclamate was injected into eggs. The warning against shoe fluoroscopy became public through ordinary news reports based on two articles in the *New England Journal of Medicine*; it is coded "routine."

Sometimes it is difficult to judge whether coverage is extraordinary or exaggerated. Consider the warning arising from the botulism death of a New York banker on June 30, 1971, and the illness of his wife a day after both ate Bon Vivant vichyssoise. The FDA ordered a recall of all the company's soups, including those distributed under 34 other labels. Of the first 324 cans of vichyssoise tested, five were contaminated with botulin, although no more illness was linked to the company.

Lawless notes, "The incident was widely publicized by the news media and created much public alarm. The *New York Times*, for example, had 27 news entries on the subject during July 1971, and several articles on botulism appeared during the same month in popular magazines [including *Time* and *Newsweek*, followed by a report in *Life* two months later]" (1977, 106). That summer the FDA recalled two of the companies' other products because of suspected botulism. One instance was confirmed, the other was a false alarm never officially acknowledged by the agency. The

FDA was suffering public criticism at the time (see Chapter 3). Its officials may have felt vulnerable on learning there had been no FDA inspection of the Bon Vivant plant in the previous four years, although yearly inspections were the agency's goal (Lawless 1977).

It is difficult in this case to know what an ordinary amount of coverage would be. The owners of Bon Vivant, which went bankrupt, asserted that the tragic error was limited to a few cans, and that the FDA was unduly hasty in condemning all its products. Yet one person had died and others—if not warned—might have. Faced with this ambiguity, I depended on Lawless's assertion that the story was "widely publicized by the news media and created much public alarm" to code it "hyped."

MEDIA SCORE

One would expect warnings hyped in the press to have received the most media coverage; indeed, one might accept this as an alternate definition of the term. Lawless provides a quantitative measure of media coverage for each case. He calculates this score as the sum of the number of entries in the *Readers' Guide to Periodical Literature*, *Business Periodicals Index*, and the *New York Times Index*, plus 10 percent of the total column inches in nine selected magazines. At the time of Lawless's calculation, recent warnings had fewer years of news coverage than older warnings, potentially confounding the media score. Yet there is only a modest correlation ($r = -.34$) between media score and the year a warning first became public.

Table 6-1 shows hyped and routine warnings in descending order of media score. Warnings I coded as hyped, on qualitative grounds, generally received higher media scores from Lawless than those I coded as routine, corroborating most of my designations.

WAS WARNING DERIVED?

Lawless comments that at least half the alarms in his sample were preceded by an earlier alarm over a related technology. "The existence of a previously related case apparently has a great sensitizing effect on the

Table 6-1. Lawless's Media Scores for Hyped and Routine Warnings

Case	*Media score*
1. DDT	**1,236**
2. Fluoridation	**968**
3. Salk polio vaccine	**326**
4. Thalidomide	**261**
5. Nuclear test on Amchitka	**247**
6. Cyclamate	**230**
7. Contaminated cranberries	**196**
8. Medical x rays	**173**
9. DES in livestock	**152**
10. Nerve gas disposal	**118**
11. Botulism	**115**
12. Oral contraceptives	**100**
13. Mercury pollution from industry	**95**
14. MSG	**92**
15. Mercury in tuna	**87**
16. Fish protein concentrate	79
17. Asbestos	72
18. Chemical Mace	**67**
19. Poison gas released at Dugway	56
20. Enzyme detergents	**48**
21. Hexachlorophene	48
22. NTA in detergent	44
23. Smog in Donora	37
24. Radiation from defective televisions	28
25. Shoe fluoroscopes	28
26. DMSO	28
27. Injuries on synthetic turf	26
28. Radioactive waste stored in Kansas	23
29. ELF radiation at Project Sanguine	21
30. Plutonium at Rocky Flats	10
31. Taconite pollution	9

Note: Warnings I assessed as hyped appear in bold.

Source: Lawless 1977.

Table 6-2. Assessment of Lawless Warnings as Derived or Pristine

	Case	*Lawless's determination of whether an alarm was preceded by earlier alarms*	*Issues that preceded alarm, if any, according to Lawless*	*Others issues that preceded alarm, according to Mazur*	*Assessment of warning*
1.	Oral contraceptives	Yes	Thalidomide, DDT	Women's movement	Derived
2.	Contaminated cranberries	No	Delaney	Delaney and DES	Derived
3.	DES in livestock	Yes	Delaney	–	Derived
4.	Cyclamate	Yes	Thalidomide, Delaney	–	Derived
5.	MSG	Yes	Cyclamate	–	Derived
6.	Botulism	No	–	–	Pristine
7.	Fish protein	No	–	–	Pristine
8.	Fluoridation	No	–	Fear of socialism	Derived
9.	Salk poio vaccine	No	–	–	Pristine
10.	Thalidomide	No	–	–	Pristine
11.	Hexachlorophene	Yes	Cyclamate, MSG, mercury in tuna	–	Derived
12.	DMSO	No	Thalidomide	Thalidomide	Derived
13.	Shoe fluoroscopes	Yes	–	Cannot identify a precursor	Pristine
14.	Medical x rays	Yes	Fallout	–	Derived
15.	Radiation from defective televisions	Yes	Fallout, medical x rays	–	Derived
16.	Smog in Donora	Yes	–	Cannot identify a precursor	Pristine
17.	Mercury pollution in industry	Yes	DDT	Lake Erie eutrophication	Derived

Table 6-2. Assessment of Lawless Warnings as Derived or Pristine *(continued)*

Case	*Lawless's determination of whether an alarm was preceded by earlier alarms*	*Issues that preceded alarm, if any, according to Lawless*	*Others issues that preceded alarm, according to Mazur*	*Assessment of warning*
18. Mercury in tuna	Yes	Mercury in industry	–	Derived
19. DDT	No	Thalidomide, cranberries	These and fallout	Derived
20. Asbestos	Yes	Mercury in industry	–	Derived
21. Taconite pollution	Yes	Lake Erie, DDT	–	Derived
22. Enzyme detergents	Yes	Water pollution, detergents	Lake Erie eutrophication	Derived
23. NTA in detergents	Yes	Lake Erie	Other detergents	Derived
24. Plutonium at Rocky Flats	Yes	War in Vietnam, nuclear power	–	Derived
25. Radioactive waste stored in Kansas	Yes	Fallout	Also nuclear power	Derived
26. Nuclear test on Amchitka	Yes	Department of Defense, antiballistic missile, environment	War in Vietnam	Derived
27. Poison gas released at Dugway	No	–	–	Pristine
28. Nerve gas disposal	Yes	Poison gas release at Dugway Proving Ground in Utah	Also war in Vietnam	Derived
29. ELF radiation at Project Sanguine	Yes	War in Vietnam, Department of Defense, environment	–	Derived
30. Chemical Mace	No	Civil rights, war in Vietnam	These are precursors	Derived
31. Injuries on synthetic turf	No	–	–	Pristine

Note: A dash (–) means "not applicable."

Source: Lawless 1977 and author assessments.

news reporters and the public. The previous publicity [on network TV news] over the Huckleby family that was poisoned by eating mercury-treated seed grain doubtlessly increased the concern over the mercury discharges by industry, and contributed to the over-concern about trace amounts of mercury in tuna fish" (1977, 492).

Precursors were not a major concern of Lawless, but he briefly notes in an appendix if each warning was preceded by a closely related or similar alarm. These ratings, coded "yes" or "no," are reproduced in the first data column of Table 6-2. Lawless again broaches the subject of precursors in discussing individual cases, noting when a warning derived from other issues in the news. These comments, summarized in the next column of Table 6-2, are usually consistent with his yes/no codes. I regarded a warning as derived if Lawless codes it "yes" and lists valid preceding issues. I regarded a warning as pristine (not derived) if Lawless coded it "no" and identified no preceding issues (Table 6-2).

Lawless codes four cases "no"—cranberries, DMSO, DDT, and mace—but identifies preceding issues. I cannot account for these inconsistencies; possibly they are coding errors. I gave weight to his specification of valid preceding issues, counting these four cases as derived. Lawless codes two cases—shoe fluoroscopes and the smog episode at Donora —as derived, but cites no preceding issue. Since I could not identify a plausible precursor for either case, I counted those as pristine warnings. Finally, I overrode Lawless on fluoridation, a controversy I have studied for three decades, counting it as a derived warning because early opposition was explicitly embedded in fears prominent in the 1950s of domestic communism and socialized medicine (Appendix II). My amendments are listed in the next-to-last column of the table.

Now, with these essential details out of the way, we may proceed to results. In Chapter 7, I rely on these codings to search for hallmarks that predict whether warnings will prove true or false.

CHAPTER 7

HALLMARKS OF TRUE AND FALSE ALARMS

Can early clues help us distinguish true warnings from false alarms? Did warnings of the 1950s and 1960s that proved valid have a different look from claims that came to nothing? Using Lawless's cases, coded according to the attributes described in the previous chapter, we can answer these questions and evaluate the hypotheses raised in Chapter 1.

IS A WARNING FIRST RAISED IN A SCIENTIFIC PUBLICATION MORE LIKELY TO BE TRUE?

Lawless found barely a warning to which some scientist or physician had not lent support, if not as initiator then as later endorser. The mere presence of legitimate scientific credentials is therefore insufficient to indicate whether a threat is real or not.

A better clue is the journalist's source—scientific, governmental, or citizen-activist—for the first public report of the threat. Table 7-1 lists true and false warnings by type of initial news source. (Case 15, on defective television sets, is excluded because the primary news source was corporate.) When the major news source was scientific, 90% of alleged threats (9 out of 10) turned out to be true. When government officials were the major source, 42% (5 out of 12) turned out to be true. When citizen activists were the major source, 38% (3 out of 8) turned out to be true (p = .03).[1] Thus, warnings from scientific sources are more than twice as likely to be true as warnings from government or citizen sources.

In Table 7-1, scientific warnings are further categorized according to whether they are based directly on peer-reviewed publications. All six peer-reviewed warnings are true; three of four non–peer-reviewed warnings are true. With so few cases, the slight difference is not meaningful.

Table 7-1. True and False Warnings, by Initial News Source

Source of first news reports	*False alarm*	*True warning*
Citizen	5. MSG	19. DDT
	8. Fluoridation	24. Plutonium at Rocky Flats
	22. Enzyme detergents	27. Poison gas released at Dugway
	26. Nuclear test on Amchitka	
	30. Chemical Mace	
Government	2. Contaminated cranberries	6. Botulism
	4. Cyclamate	9. Salk polio vaccine
	7. Fish protein concentrate	16. Smog in Donora
	12. DMSO	17. Mercury pollution from industry
	18. Mercury in tuna	28. Nerve gas disposal
	21. Taconite pollution	
	29. ELF radiation at Project Sanguine	
Scientific		
Not peer reviewed	23. NTA in detergents	3. DES in livestock
		25. Radioactive waste stored in Kansas
		31. Injuries on synthetic turf
Peer reviewed		1. Oral contraceptives
		10. Thalidomide
		11. Hexachlorophene
		13. Shoe fluoroscopes
		14. Medical x rays
		20. Asbestos

If reputable science is a hallmark of true warnings, was the scientific knowledge available in the 1970s sufficient to have avoided some false alarms? Lawless suggests it was. In succinct technical assessments of his cases, he appraises if opponents exaggerated the threat, "yes" or "no." I interpret these to be reasonably objective judgments based on information publicly available in the mid-1970s. (He gives no assessments for warnings about MSG, botulism, fluoridation, or DDT.)

Table 7-2 lists true and false warnings (assessed in 2001) by Lawless's assessment as of the mid-1970s. Of the 14 threats he believed were not exaggerated, 79% are indeed true; of the 13 threats he thought were exaggerated, only 38% are true (a difference of 41 percentage points; $p = .03$). Put another way, Lawless "predicted" the 2001 assessment correctly for 19 warnings, and incorrectly for 8. Thus, the majority of Lawless's risk assessments pass the test of time, suggesting there was sufficient knowledge by

Table 7-2. True and False Warnings, by Lawless's Assessment of Threat

Lawless's assessment of threat	*False alarm*	*True warning*
Exaggerated	2. Contaminated cranberries	1. Oral contraceptives
	4. Cyclamate	3. DES in livestock
	8. Fluoridation	14. Medical x rays
	12. DMSO	15. Radiation from defective televisions
	18. Mercury in tuna	28. Nerve gas disposal
	22. Enzyme detergents	
	26. Nuclear test on Amchitka	
	30. Chemical Mace	
Not exaggerated	21. Taconite pollution	9. Salk polio vaccine
	23. NTA in detergent	10. Thalidomide
	29. ELF radiation at Project Sanguine	11. Hexachlorophene
		13. Shoe fluoroscopes
		16. Smog in Donora
		17. Mercury pollution from industry
		20. Asbestos
		24. Plutonium at Rocky Flats
		25. Radioactive waste stored in Kansas
		27. Poison gas released at Dugway
		31. Injuries on synthetic turf

the 1970s, if not earlier, to correctly appraise a majority of warnings. The implication, seemingly, is that these warnings, as a group, could have been appraised better than they were at the time.

IS PRIOR BIAS A MARK OF A FALSE ALARM?

Some people who raise alarms have preexisting biases against the corporation or government agency responsible for the alleged hazard. During the war in Vietnam, for example, warnings often came from antiwar activists blaming the Pentagon for risks imposed on civilians in the

Table 7-3. True and False Warnings, by Pre-existing Bias toward Party Responsible for Alleged Threat

Was there prior bias?	*False alarm*	*True warning*
Yes	5. MSG	3. DES in livestock
	8. Fluoridation	19. DDT
	12. DMSO	24. Plutonium at Rocky Flats
	21. Taconite pollution	28. Nerve gas disposal
	22. Enzyme detergents	
	26. Nuclear test on Amchitka	
	29. ELF radiation at Project Sanguine	
	30. Chemical Mace	
No	2. Contaminated cranberries	1. Oral contraceptives
	4. Cyclamate	6. Botulism
	7. Fish protein concentrate	9. Salk polio vaccine
	18. Mercury in tuna	10. Thalidomide
	23. NTA in detergents	11. Hexachlorophene
		13. Shoe fluoroscopes
		14. Medical x rays
		15. Radiation from defective televisions
		16. Smog in Donora
		17. Mercury pollution from industry
		20. Asbestos
		25. Radioactive waste stored in Kansas
		27. Poison gas released at Dugway
		31. Injuries on synthetic turf

United States by military-industrial technologies. Are alarms coming from biased parties less credible than those raised by disinterested parties?

Table 7-3 compares warnings raised by an identifiably prebiased party with warnings raised by an apparently disinterested source. Of 19 warnings with no identifiably negative bias, 74% are true; of 12 warnings raised by someone with a prior bias, 33% are true (a difference of 41 percentage points; $p = .09$). The presence of prior negative bias somewhat increases the odds that an alarm is false.

Before concluding too quickly that bias compromises a warning, we must consider that 9 of 10 scientific warnings are coded as unbiased. (This is not surprising since orthodox scientific institutions wear a veneer of objectivity, so prejudices—if they do exist—are not easily visible to the public.) One might argue that unbiased warnings tend to be valid because they usually come from scientific sources, not because they are unbiased per se. To evaluate this possibility, Table 7-4 is limited to warnings from nonscientific sources, thus eliminating the conflation of bias with scientific origin. Of 10 nonscientific warnings with no bias, 60% are true; of 11 nonscientific warnings raised by a party with an identifiable bias, 27% are

Table 7-4. True and False Warnings, by Pre-existing Bias toward Party Responsible for Alleged Threat: Nonscientific Sources Only

Was there prior bias?	*False alarm*	*True warning*
Yes	5. MSG	19. DDT
	8. Fluoridation	24. Plutonium at Rocky Flats
	12. DMSO	28. Nerve gas disposal
	21. Taconite pollution	
	22. Enzyme detergents	
	26. Nuclear test on Amchitka	
	29. ELF radiation at Project Sanguine	
	30. Chemical Mace	
No	2. Contaminated cranberries	6. Botulism
	4. Cyclamate	9. Salk polio vaccine
	7. Fish protein concentrate	15. Radiation from defective televisions
	18. Mercury in tuna	16. Smog in Donora
		17. Mercury pollution from industry
		27. Poison gas released at Dugway

true (a 33% difference, $p = .09$). Thus, prior negative bias continues to increase the odds that a warning is false independently of the source of the warning.

IS HYPED MEDIA COVERAGE A MARK OF A FALSE ALARM?

Table 7-5 compares true and false warnings when press coverage was hyped versus when it was routine (see also Chapter 6, p. 83). Of 17 warnings hyped by promoters, 53% are true. Of 14 warnings covered in a routine way, 64% are true (a difference of 11 percentage points; not statistically significant). Contrary to the hypothesis, hype provides no clue to the validity of a warning.

Table 7-5. True and False Warnings, by Press Coverage Hyped or Routine

Type of press coverage	*False alarm*	*True warning*
Hyped	2. Contaminated cranberries	1. Oral contraceptives
	4. Cyclamate	3. DES in livestock
	5. MSG	6. Botulism
	8. Fluoridation	9. Salk polio vaccine
	18. Mercury in tuna	10. Thalidomide
	22. Enzyme detergents	14. Medical x rays
	26. Nuclear test on Amchitka	17. Mercury pollution from industry
	30. Chemical Mace	19. DDT
		28. Nerve gas disposal
Routine	7. Fish protein concentrate	11. Hexachlorophene
	12. DMSO	13. Shoe fluoroscopes
	21. Taconite pollution	15. Radiation from defective televisions
	23. NTA in detergents	16. Smog in Donora
	29. ELF radiation at Project Sanguine	20. Asbestos
		24. Plutonium at Rocky Flats
		25. Radioactive waste stored in Kansas
		27. Poison gas released at Dugway
		31. Injuries on synthetic turf

DOES A DERIVED WARNING HINT AT A FALSE ALARM?

A hazard may become temporarily newsworthy for diverse reasons. The actual severity of the risk—the body count—is one among several factors affecting an editor's decision to run a story. Also important is the hazard's relevance to other stories currently or recently reported, or to major issues of concern at the time. During the years of protest against the war in Vietnam, hazards of military origin, even those having no direct connection to the war, were unusually newsworthy (Chapter 3). Analogously, in the months after a catastrophic accident, "near misses" that one ordinarily never hears about are reported (Mazur 1990; Stallings 1994). The Y2K glitch was the most prominent of several vacuous millennium-related warnings in the news as we approached the year 2000.

Table 7-6. True and False Warnings, by Whether the Warning Is Derived or Pristine

Type of warning	*False alarm*	*True warning*
Derived	2. Contaminated cranberries	1. Oral contraceptives
	4. Cyclamate	3. DES in livestock
	5. MSG	11. Hexachlorophene
	8. Fluoridation	14. Medical x rays
	12. DMSO	15. Radiation from defective televisions
	18. Mercury in tuna	
	21. Taconite pollution	17. Mercury pollution from industry
	22. Enzyme detergents	
	23. NTA in detergents	19. DDT
	26. Nuclear test at Amchitka	20. Asbestos
	29. ELF radiation at Project Sanguine	24. Plutonium at Rocky Flats
		25. Radioactive waste stored in Kansas
	30. Chemical Mace	
		28. Nerve gas disposal
Pristine	7. Fish protein concentrate	6. Botulism
		9. Salk vaccine
		10. Thalidomide
		13. Shoe fluoroscopes
		16. Smog in Donora
		27. Poison gas released at Dugway
		31. Injuries on synthetic turf

When a hazard enters the news simply because it complements another news story, we may wonder if it is really all that hazardous. A pristine warning, one making the front page on its merits alone, without a boost from collateral issues, may deserve more respect. Do derived warnings hint at false alarms?

Table 7-6 lists true and false warnings by whether they are derived or pristine. Of eight pristine warnings, 88% are true; of 23 derived warnings, 48% are true (a difference of 40 percentage points; $p = .05$). Indeed, pristine warnings are valid more often than derived warnings.

We have already seen that nearly all warnings from scientific sources turned out to be true, whether derived or pristine. The difference in validity between derived and pristine alarms must therefore be greatest among warnings raised by nonscientific news sources. We may check this in Table 7-7, which again compares derived versus pristine warnings, but only for those raised by nonscientific news sources. Of 5 pristine nonscientific warnings, 80% are true; of 16 derived nonscientific warnings, 31%

Table 7-7. True and False Warnings, by Whether the Warning Is Derived or Pristine: Nonscientific Sources Only

Type of warning	*False alarm*	*True warning*
Derived	2. Contaminated cranberries	15. Radiation from defective televisions
	4. Cyclamate	17. Mercury pollution from industry
	5. MSG	19. DDT
	8. Fluoridation	24. Plutonium at Rocky Flats
	12. DMSO	28. Nerve gas disposal
	18. Mercury in tuna	
	21. Taconite pollution	
	22. Enzyme detergents	
	26. Nuclear test on Amchitka	
	29. ELF radiation at Project Sanguine	
	30. Chemical Mace	
Pristine	7. Fish protein concentate	6. Botulism
		9. Salk polio vaccine
		16. Smog in Donora
		27. Poison gas released at Dugway

are true (a difference of 49 percentage points; $p = .05$). It is among these nonscientific warnings—brought to the news by government officials or citizen activists—that pristine warnings are more likely than derived warnings to be true.

MULTIVARIATE ANALYSIS

The clues to valid warnings discovered in this chapter may be redundant, as noted in the overlap between bias and type of news source. It is also conceivable that some relationships shown here are spurious, capable of being explained away by differences between older and more recent warnings. I tested these possibilities using logistic regression analysis, which estimates the effects of multiple independent variables on a dichotomous dependent variable. The dependent variable, the validity of a warning, is coded 1 if true, 0 if false.

I compared various combinations of three independent variables: type of news source (coded as a dichotomy: 1 if scientific, 0 if government or citizen), whether the warning is derived or pristine (coded 0/1), and the year of the first public warning. Hype is excluded since it was shown to be unrelated to the validity of a warning (Table 7-5). Prior bias is excluded because of its strong colinearity with the type of news source. The coefficient for the year of first warning is approximately zero in all models, meaning that warnings raised in the early postwar years were no more or less valid than warnings raised later.

The coefficient for type of news source remains in the narrow range of 2.5–2.7 (always $p = .03$), whether it is the sole independent variable in a model or used in combination with one or both of the others. Thus, the tendency of warnings to be valid if they come from scientific news sources is unaffected by controlling for derived-versus-pristine warning or year of warning.

The coefficient for derived-versus-pristine warning also remains in a narrow if slightly lower range of 2.0 ($p = .10$) to 2.3 ($p = .07$), whether it is the sole independent variable or used in combination with other variables. Thus, the tendency of warnings to be valid if they are pristine is unaffected by controlling for type of news source or year of warning.

WHAT THE ANALYSIS REVEALS

Usually from their earliest moments, valid warnings in Lawless's sample looked different from those eventually judged mistaken. Alarms more often turned out to be true when their news source was a report of normal scientific research produced at a recognized scientific institution than when the source was a government agent or citizen advocacy group. Warnings reaching the news from unbiased sources were more often true than warnings coming from identifiably prejudicial sources. Derived warnings (those appearing in the news partly because of their connection to earlier news stories) were more often false than pristine warnings reaching the news without a boost from collateral stories. Hyped warnings, those for which sources or journalists made unusual efforts to increase news coverage, were no more or less likely to be valid than warnings given routine treatment by the media.

Note

1. Because Lawless's cases are not a random sample of any well-defined universe of warnings, they lack the prerequisite for proper application of a significance test. I report significance levels because readers often want them even when they are not technically valid. These are based on chi-squared tests with continuity correction for small cell frequencies.

CHAPTER 8

HINDSIGHT AND FORESIGHT

The clearest hallmark of a true public warning during the period 1948–1971 was a reputable scientific news source. Warnings reaching the press from scientists operating in a conventional way at an orthodox scientific institution were true more than twice as often as those reaching the news from government officials or citizen advocates.

Before, say, 1960, science was often seen as a completely objective pursuit. Then it would have been a foregone conclusion that sound research ensures valid risk assessment. Today we recognize old claims of total objectivity as a rhetorical veneer that legitimated scientific viewpoints. A few social theorists go further, denying that scientific knowledge is in any way superior to knowledge produced by other forms of inquiry. But the presence of subjective influences does not imply that science has no credibility at all. In Lawless's era, at least, reputable scientific warnings did have greater validity than alarms coming from other sources.

A discouraging finding is that warnings originating with government agencies or officials were, in retrospect, more often false than true (Table

7-1). Involved agencies, usually the FDA, had scientists on staff but were essentially concerned with policy and administration, not research. Government issuance and retraction of alarms, such as those concerning cranberries, cyclamate, and mercury in tuna, often seemed overly responsive to political considerations.

My analysis of Lawless's cases revealed additional hallmarks of a true warning. Pristine alarms were more often true than derived warnings riding the coattails of collateral news stories. Also, warnings coming from unprejudiced sources were more often true than those raised by parties with prior biases. However, hyped warnings were no more or less valid than those receiving less-sensationalized press coverage.

IS THE STUDY VALID, AND DO THE HALLMARKS APPLY TO MORE RECENT WARNINGS?

Social scientists traditionally evaluate a study in two ways: Is it *internally* valid—in this case, are my findings believable for the 31 cases under analysis? If so, is the study *externally* valid—do the findings for these 31 cases generalize to a broader class of public warnings?

I discuss safeguards of internal validity in Chapters 5 and 6, and provide the reasons for all coding decisions in Appendix I. Anyone can revisit any case using mostly public documents. I checked Lawless's work on each case, and found few factual or documentary errors; I believe his codes and judgments are accurate and fair. I also submitted my retrospective assessments of the warnings as true or false to several reviewers with diverse perspectives. After I addressed their criticisms, they judged my assessments as mostly but not entirely acceptable. I could not satisfy one reviewer with a strong environmental health orientation who insisted that the fluoridation warning is true, and that hardly any warning should be regarded as a false alarm since almost none could be entirely discounted

Do my findings generalize beyond Lawless's 31 cases, and beyond the period 1948 to 1971? Conclusions about external validity are necessarily inductive and therefore never definitive. Moreover, the way we handle risks has changed greatly since the years of study. The U.S. EPA did not exist prior to 1970. Today's nongovernmental environmental organizations have more technical expertise than did citizen activists of the postwar period, promising more credibility as sources of warnings. Academic scientists, increasingly dependent on biotechnology companies and other industrial sources of funding, have conceivably become less credible (Krimsky 1991).

The risk landscape has changed continually since World War II. The 1950s were as different from the 1960s as the 1960s were from the 1970s. Nuclear weapons, both their testing and their use, were the dominant fear of the 1950s and early 1960s, a time of little concern or even knowledge about chemical carcinogens; by the early 1970s these concerns were inverted. The modern environmental movement was born in the 1960s when public trust in governmental institutions was dissolving, and when the ability of chemists to measure trace contaminants in food and the environment was rapidly improving. On that basis one could question whether findings from one decade are inapplicable to the next, essentially denying any value to historical analysis. Rather than adopting this extreme position, we can examine in more detail the relevant changes that have occurred since Lawless stopped collecting data.

Warnings in the 1970s

During the 1970s, just after Lawless's era, the frequency and stridency of technological and environmental warnings (and countering assurances) became so cacophonous as to strain the credibility of fair-minded observers. Walter Sullivan, the eminent science writer for the *New York Times*, told me he had passed up the 1974 story that human-made chlorofluorocarbons could deplete the ozone layer because there was so much "doomsday reporting" going on, and he did not want to publicize too quickly the latest prediction of environmental disaster. Of course, this turned out to be a true warning.

The decade also saw warnings over the ABM, the supersonic transport, nuclear power plants, recombinant DNA, and toxic waste (Mazur 1981, 1998). The frequent expression in the public arena of contradictory factual claims by scientific adversaries fed a growing distrust of scientific institutions, even of charges of bias against the previously august National Academy of Sciences (e.g., Boffey 1975, Schiefelbein 1979). Scientist jokes approached (but never attained) a level known by the legal profession. A "science court" was proposed to separate credulous from noncredulous warnings, then itself became the focus of heated controversy (see *Risk: Issues in Health & Safety*, volume 4, spring 1993, for a retrospective on the Science Court).

Lacking a systematic survey of public warnings during the 1970s, we are easily drawn to that decade's most salient if not most representative examples: Three Mile Island and Love Canal. The reactor accident in 1979 was a stupendous media event, far surpassing coverage devoted two months later to the nation's worst airline accident, the crash of a DC-10 that killed 274 people because of a faulty engine mount assembly. The Three Mile Island disaster killed no one but occurred after years of assurances by the nuclear industry and the Nuclear Regulatory Commission (NRC) that such mishaps were virtually impossible (although accidents beginning similarly had occurred before; Mazur 1984). By a quirk of timing, the accident came just days after an antinuclear movie, *The China Syndrome*, starring Jane Fonda and Jack Lemon, was released to theaters across the nation. Initial news accounts juxtaposed the movie's fictional accident with the real one, but as the accident developed real events became more dramatic, reaching a climax when NRC officials warned that a hydrogen bubble in the Three Mile Island reactor might explode, releasing radioactivity to the surroundings. In fact, there was insufficient oxygen in the reactor to support an explosion. The warning was quickly withdrawn, but it was too late to prevent panic near the site (President's Commission 1979; Sandman and Paden 1979).

Love Canal was a two-year episode, lasting from the first public warning of toxic chemicals seeping into homes near the abandoned dumpsite to the final evacuation of the neighborhood. The controversy overlaps Three Mile Island in time and matches it as a symbol of technological disaster. Here citizens campaigning for a government buyout of their homes gave journalists information rife with error and bias, portraying their community as suffering widespread chemically induced illness. The buyout was finally accomplished when an unquestionably inadequate study by the EPA seemed to show residents suffering an unusual degree of chromosome breakage, results not sustained when subjected to proper research methods (Mazur 1998).

No informed observer denies that these sites posed real hazards. The reactor core at Three Mile Island partially melted; hazardous chemicals reached basements of some homes immediately bordering the Love Canal. If alarms raised by some government and citizen sources were exaggerated, reassurances issued by other officials and corporate spokespeople were off the mark as well. In each case, contemporary news reports failed to provide consistent or, in retrospect, correct appraisals of risk.

Warnings since 1980

Love Canal and Three Mile Island were public relations disasters for the chemical and nuclear industries and for government agencies charged with oversight. Coming atop criticism during the 1970s about policy based on unsound science, they compelled reforms during the 1980s in rhetoric and perhaps in practice, too.

The nuclear industry's determination to put its house in order, though failing to revive nuclear energy in America, deserves credit for the post-1979 safety record of U.S. reactors. Spurred by publicity about Love Canal, Congress in 1980 enacted the Superfund law requiring the cleanup of abandoned hazardous waste sites. Superfund was soon amended to require facilities that manufacture, process, or use toxic chemicals to document their waste flows of these materials. Yearly publication of this information in the EPA's Toxic Release Inventory fed annual news stories about each city's major industrial polluters, inspiring some corporations to voluntarily cut their waste to avoid further condemnation. Chemical companies had by this time lowered their aspirations for public esteem, no longer boasting, as Dupont did in the 1950s and 1960s, of "better things for better living through chemistry."

Government regulators—"over-regulators" in the eyes of the Reagan administration—also suffered a loss of public support, most tangibly through budget cuts and the appointment of unsympathetic administrators. When the short but famous reign (1981–1983) of President Reagan's first head of the U.S. EPA, Anne Gorsuch Burford, ended amid scandal over cronyism with industrial polluters, conservatives joined liberals in criticizing the overpoliticization of regulatory decision making, renewing calls for better science (Burford and Greenya 1986).

Quality of science is a perennial problem at the EPA, continually addressed but never solved. In the late 1990s, Mark Powell (1999), interviewing more than 30 informants familiar with EPA operations, found that perception of scientific quality varied considerably across programs. The EPA's science was seen as most sophisticated when it focused on pesticides, toxic substances, and air pollutants, and as least sophisticated when dealing with contaminated sites and hazardous wastes. Asked to compare the EPA's overall use of scientific information for regulatory decision making "currently, 10 years ago, and 20 years ago," these informants said that the agency's fair-to-poor performance in the 1970s had improved, but most still did not rate its science "very good." Whether the EPA is now a trustworthy source of warnings remains unclear.

The same question arises for the FDA, which had been embroiled in so many controversies in Lawless's era that in 1969 the incoming commissioner, Charles Edwards, complained about its "low scientific credibility" (FDA 1997). As recently as 1997, a subcommittee of the FDA's Science Board found that the agency still lacked a culture essential for nurturing high-quality science. Attempts at improvement, the subcommittee thought, had consistently failed because well-intentioned scientific aspirations were incompatible with the burden of ever-growing regulatory mandates and insufficient budgets (FDA 1997).

If doubts remain about the credibility of government regulators, what about citizen advocates? Today's major environmental organizations, much concerned with the kinds of hazards Lawless reports, barely existed in 1970, or, in the case of old-line conservation groups, they existed in very different form. Some, notably the Union of Concerned Scientists (founded in 1969) and the National Resources Defense Council (NRDC, founded in 1970), have staff with impressive scientific credentials. Still, these organizations operate in a primarily polemical mode, openly prejudiced against alleged purveyors of public hazard.

Like industrialists and government regulators, environmentalists suffered their own public relations Waterloo over Alar, not as outrageous as Three Mile Island or Love Canal but nonetheless damaging to public trust. While associates of NRDC still defend its 1989 warning about Alar—a chemical used by growers to delay the ripening of apples to improve their marketability—those associates also acknowledge the widespread perception of the episode as a false alarm. The warning's validity depends on whether tumors produced in lab animals fed very large amounts of a chemical indicate carcinogenicity at trace doses (though by any reasonable reading the cancer hazard from Alar is not large). Most damaging to the NRDC's position, probably, were reports by critics that highlighted the organization's backstage efforts to hype the Alar warning, which won the warning sympathetic coverage on television's *60 Minutes* and helped enlist actress Meryl Streep as a spokeswoman for the warning (Gots 1993). Deservedly or not, Alar to this day is waved as a red flag against environmental hyperbole (Linda Greer, NRDC, pers. comm.).

This review of past decades suggests that the hallmarks I identified as characteristic of warnings from 1948 to1971 still carry weight. A caveat is that new, scientifically sophisticated environmental groups might have valid warning capacity if they escape capture by political prejudices. More recent cases must be evaluated to determine whether they in fact fulfill this role.

In the meantime it seems prudent to trust warnings based on reputable science that is relatively free of prior biases. An obvious case in point is the warning about human-induced global warming, assessed periodically by the Intergovernmental Panel on Climate Change, a U.N. body commissioned to appraise current scientific understanding and advise the world's governments on this hazard (IPCC 2001). The consensus on global warming is not total and never will be. Complete agreement within the scientific community is like the end of the rainbow, approachable but never reachable. Yet agreement is close enough that we ought to believe the warning, whether or not we want to do anything about it.

HOW THE NEWS MEDIA INFLUENCE WARNINGS

Truth, for most adults, comes from the news media, which throughout the postwar era have been criticized—sometimes deservedly, sometimes not—for inaccurate and biased reporting about technological risks (President's Commission 1979; Singer and Endreny 1994; Lichter and Rothman 1999). A related problem is the ready acceptance by some journalists of risk information from biased or incompetent sources, even occasionally from children (e.g., Mazur 1998). Another problem is the tendency of news editors and producers to publicize risk warnings that make interesting reading or viewing, whether or not they are credible. Warnings may be reported for no better reason than their relationship to other ongoing stories, reaching the front page or network television news by riding the coattails of collateral events. We have seen here that such derived warnings are less likely to turn out to be true than pristine warnings reaching the news on their own merits.

Perhaps more important than the link between derivation and validity is the large number of public alarms that are derived from other stories. Among Lawless's 31 warnings, 74% had roots in earlier publicized concerns (Table 6-2). A genealogy of warnings, charted in Figure 8-1, shows the kinship of 15 chemical hazards (in capital letters). Three nodes of the diagram (in lower case) represent additional news stories Lawless cites as progenitors. These are congressional hearings leading to passage in 1958 of the Delaney Clause banning food additives shown to cause cancer in any experimental animal; the controversy over fallout from atmospheric testing of nuclear weapons; and the pollution-induced "death" of Lake Erie from eutrophication. Each solid arrow corresponds to a comment by Lawless that an earlier warning or issue influenced a later warning. Dotted arrows show three additional connections not mentioned by Lawless but apparent to me.

This genealogy, besides elucidating family ties among alarms, suggests the degree to which the appearance of a warning in the news depends on its relationships to earlier news stories. Orphan warnings have a tougher time breaking into the media. Does this mean that we are missing important problems not obviously connected to current public concerns?

Journalists should search not only more widely but more deeply for true warnings, to determine if there is a serious basis for concern. It is tiresome for critics—and unfair to good reporters—to harangue the press and television for baseless scare stories serving no purpose but entertainment. The United States, unlike Canada, has no law against reporting false news; our First Amendment protections are too precious to even hint at press censorship. The press guards its own practices, sometimes rigorously. A peculiar instance was the World War II agreement among American journalists to avoid showing President Roosevelt as crippled. Today the press

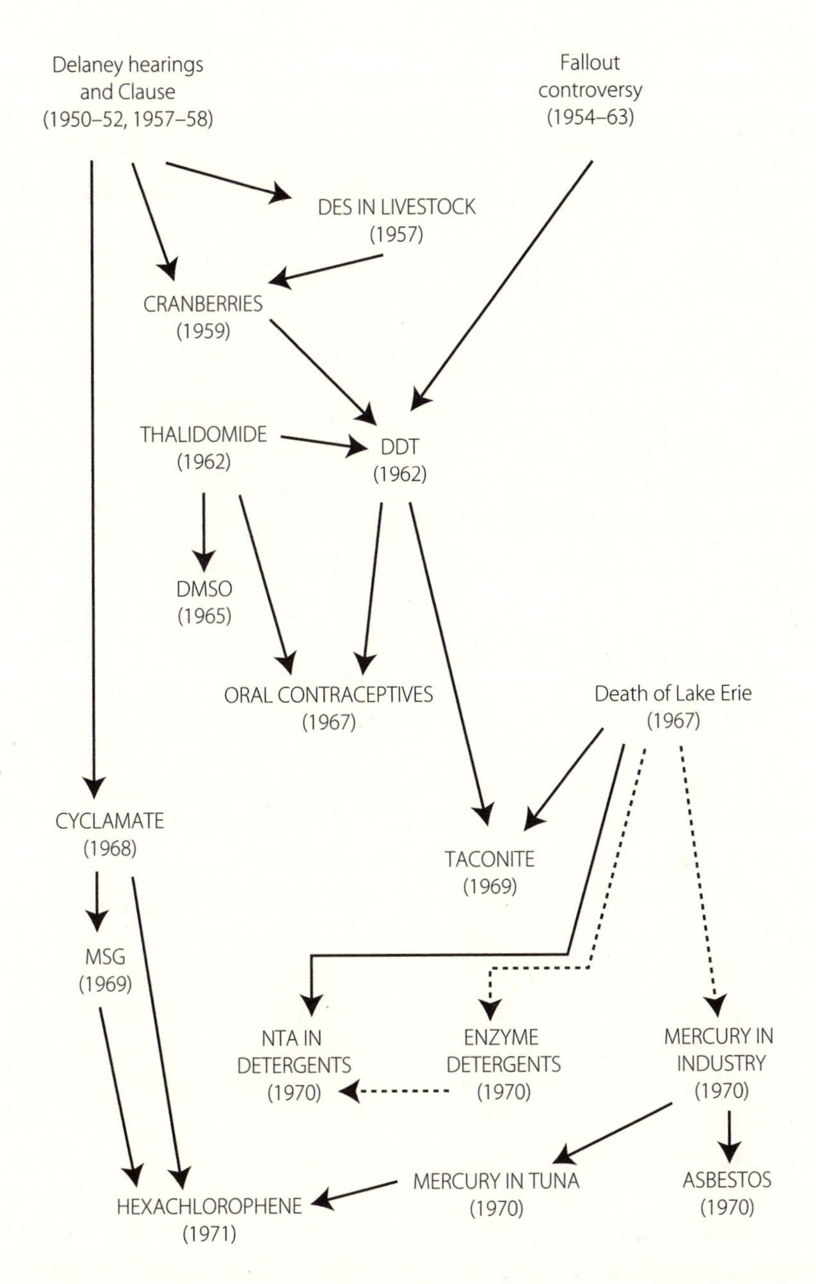

Figure 8-1. Lawless's Genealogy of Chemical Warnings (Dotted Lines Indicate Connections Made by Mazur)

self-enforces bans on language and visuals that violate common standards of decency; on insults to minority groups; and on identification of confidential sources even when such a ban denies accused parties knowledge of their accusers. Journalists exercise admirable responsibility in such matters. Perhaps they could add to that list a ban on overtly insubstantial false alarms.

ARE FALSE ALARMS UNWISE?

Before concluding this study, it is worth emphasizing that a warning shown in retrospect to be false may still have been wise or justified, from some perspective, at the time it was raised. Ralph Nader harangued baby food manufacturers for adding monosodium glutamate (MSG) to their products, claiming that it might cause brain damage in infants. The manufacturers responded, correctly as it turns out, that MSG is harmless and generally irrelevant to infants. Still, it is easy to side with Nader in this case, even if the basis for his warning was meager. The chemical, along with salt, was put into baby food to make it tastier to mothers who purchased the product; MSG had no benefit for infants. What besides profit was the justification for exposing babies to even an unlikely hypothetical risk?

A warning eventually recognized as incorrect might be wise at the time if the consequences of inaction are potentially severe. A suspected carcinogen, rapidly spreading through the population at high exposure levels, deserves a serious warning even if compelling evidence is not in hand. The slower the spread and the lower the exposure to a potential toxin, the less need for quick action and the more time to weigh evidence. Perhaps because exposure to contaminated Thanksgiving cranberries would have been so low and so brief, no one retrospectively defends that warning as either true or justified.

Warnings that are false may still be useful and therefore wise for advancing political goals. The alarm raised over an underground nuclear weapons test on Amchitka Island, though never rooted in serious scientific concerns, was effective as part of the larger protest against the war in Vietnam. The warning about the Reserve Mining Company's taconite tailings, incorrect as a health threat, led the company to cut its pollution of

Lake Superior. Communities across the nation continue to hear false warnings about fluoridation. As a result, they usually reject the water treatment when it is the subject of a referendum, producing a victory for antifluoridationists who object to government-imposed health measures. The false health warning about Mace was an effective retort to police brutality during civil rights demonstrations and the 1968 Democratic convention in Chicago. A warning's eventual truth or falsity clearly has little to do with whether it is "wise" from someone's perspective.

From the viewpoint of a technological optimist, any dubious warning can be unwise if it thwarts innovation. The optimist's guideline might be the Principle of Prudent Progress, urging us forward with beneficial new technology while monitoring for damage along the way. The pessimist's guide is the Precautionary Principle, urging restraint on technological innovation in the presence of any plausible warning. Both are platitudes, better as political slogans than as serious principles for public policy.

DOES THE TRUTH MATTER?

In a democracy, the people or their representatives are free to spend public money as they see fit. Interest groups compete to channel funding to their favorite causes. If U.S. society chooses to allot far more money to cleaning up toxic waste sites, which harm few people, than to prevent teenagers from smoking, which creates an enormous health burden, that is our privilege as a nation.

Still, many risk analysts are disturbed when we fail to maximize the number of lives saved per dollar spent on risk remediation. They point out that actions taken by government to avoid the consequences of an alleged hazard are often unrelated to the severity or scientific validity of the hazard (EPA 1982; Breyer 1993; Graham and Wiener 1995; Mazur 1998). The inference is that policy should be better aligned with science, and that irrational or inefficient elements of policymaking should be eliminated (but see Mazur 1985 and Driesen 2001 for limitations on this position).

Yet public policy does not follow directly from scientific knowledge. A value-laden subjective decision is always involved, one that requires weighing pros and cons, costs and benefits, winners and losers. A wise policy choice for one party with certain interests may not be the wisest choice for a party with different interests. These considerations raise a question: does scientific evaluation of a warning matter at all?

Essentially two models show how science is applied to public policy. The first—call it the "knowledge model"—assumes that scientists can obtain approximately true answers to their research questions with methods that are fairly objective. This knowledge is used to inform public policy. For example, scientists can determine the health risk from exposure to fluoride at levels adequate to prevent cavities. Policymakers then use this finding as one factor in deciding whether to add fluoride to community drinking water. Such decisions cannot follow from facts alone, but facts ought to influence outcomes. If the health risk is high, that should help shift the decision against fluoridation; if low, that should encourage fluoridation. This model makes no sense to anyone who denies that science can find correct answers.

The second model—the "politics model"—can be applied whatever one's view concerning the objectivity of science. Here partisans use scientific findings as political capital to sway policy in the direction they prefer. If such partisans favor fluoridation, they claim there is little health risk; if they oppose it, they claim a high health risk. It makes no difference if findings are correct, objective, or honest as long as they are persuasive. These actors bury findings that work against their position, or attack them as invalid or inapplicable. In the politics model, scientific claims are used polemically, just like any other kind of political argumentation (Mazur 1998; Brown 1991).

The politics model has many proponents. Partisans in a particular controversy often see their goal as sufficiently important to justify any interpretation of scientific data that is favorable to their case. During breaks from writing this final chapter, I am reading John McPhee's (1971) laudatory biography of David Brower, a major environmentalist of the postwar period. McPhee repeatedly describes Brower's habit of making up "facts" to support his arguments against industrialists and developers. The biographer seems to regard this as an endearing tactic of the "archdruid" in his advocacy for wilderness preservation. Like McPhee, we sympathize with those who fight the good fight, accepting their argumentation when in other contexts it would be vexing.

But the politics model loses its appeal if applied to the entire array of technical controversies affecting policy. Science that is sufficiently malleable to serve any position in one controversy can serve any position in all controversies, and in that event science does not matter at all. The famous parable of "the tragedy of the commons" tells how each shepherd maximized his own herd's grazing on the village green until no grass remained for anyone (Hardin 1968). In the same way, if each technical expert interprets data for his or her own convenience, with no attempt at objectivity, there will be no experts left with unimpeachable credibility, and we will all suffer for it.

Government has limited resources to address the problems that face us. Choices are inevitable. Interest groups should carry weight in these decisions, but so should truth, to the extent that we can determine it.

APPENDIX I

SUMMARY OF CASES

Results in this book are based on 31 Lawless cases, which I summarize here. (See Chapter 6 for my rationale for choosing these cases and an explanation of how I coded them for my study.) Under "risk assessment" below, I indicate whether I judged a warning "true" or "false" based on empirical evidence (E) or regulatory standards (R) (see Chapter 5 for more explanation).

CASE 1. ORAL CONTRACEPTIVES

Nature of the warning: Birth control pills might cause stroke, blood clots, or cancer. The warning was not dramatic until Gaylord Nelson held Senate hearings in 1970.

First public warning: *New York Times*, May 23, 1967, "Doctors Say Some Women Using Birth Control Pills May Risk Stroke," p. 38.

Type of news source: Scientific, peer reviewed. On May 22, 1967, Dr. J. Gardner and two other physicians from Western Reserve University issued a recommendation that doctors use caution in prescribing the Pill based on nine users who suffered brain damage. The researchers had presented this study to the American Academy of Neurology on April 29. Apparently this work was not peer reviewed, but the *Times* also cites a peer-reviewed preliminary report from Britain's Medical Research Council, published in the *British Medical Journal* (April 27, 1967: 355), saying that women on the Pill run a higher risk of fatal strokes, heart attacks, and blood clots in the lungs.

Did source have identifiable preexisting bias? No.

Risk assessment: True(E). For comment on risk levels, see <www.fda.gov/cder/guidance/2448dft.htm>. Today's birth control pills contain lower hormone levels, reducing the risks.

Did Lawless regard this threat as overemphasized? Yes.

Media coverage: Hyped. "At [Nelson's Senate] hearings [covered on live television], critics of the Pill caused a sensation by claiming it had not been adequately tested and was the potential cause of blood clots and perhaps cancer" ... [The hearings] "generated a great many scare news stories around the country" (Lawless 1977, 28, 37).

Media score: 100.

Was warning derived? Yes, from thalidomide, DDT, cyclamate, and also the women's movement.

CASE 2. CONTAMINATED CRANBERRIES

Nature of the warning: Some cranberry growers in the Northwest used amitrole, an unapproved herbicide. Department of Health, Education, and Welfare (HEW) Secretary Flemming announced on November 9, 1959, that some cranberries contained trace residues of amitrole, which caused cancer in rats. Sales of cranberries were not banned, but Flemming suggested that housewives should not buy any. Coming 17 days before Thanksgiving, this warning sent a shock through the public and the cranberry industry.

First public warning: *New York Times,* November 10, 1959, p.1, "Some of Cranberry Crop Tainted By a Weed-Killer, U.S. Warns;" *Wall Street Journal,* November 10, 1959, p. 3, "Chemical That Causes Cancer in Animals Is Found in Some West Coast Cranberries."

Type of news source: Government. "On Monday, November 9 [1959], just seventeen days before Thanksgiving Day, [HEW Secretary Flemming] called a news conference and announced that at least two shipments of cranberries produced in Washington and Oregon in 1959 contained residues of a weed killer that could produce thyroid cancer in rats ... sale of all cranberries was not banned, Mr. Flemming said, but three million pounds of the Northwestern crop for 1957 were being buried and the 1958 and 1959 crops from all growing areas would be investigated" (Lawless 1977, 60). Studies by American Cyanamid Corporation and FDA showed rats with enlarged thyroids that were perhaps cancerous. FDA ruled that amitrole is a carcinogen. In early November, the Seattle office of the FDA reported that some cranberries were contaminated with amitrole.

Did source have identifiable preexisting bias? No.

Risk assessment: False (E). FDA counted berries with more than 0.15 ppm amitrole as "contaminated." By January 1960, 99% of the crop was cleared as uncontaminated (Lawless 1977, 64). Today amitrole is reasonably considered to be a human carcinogen based on sufficient evidence of carcinogenicity in experimental animals. The International Agency for Research on Cancer calls it possibly carcinogenic to humans. The World Health Organization (WHO) assigns amitrole an ADI of 0.0005 mg/kg/day, based on a safety factor of 1,000 (see also <ace.orst.edu/info/extoxnet/pips/amitrole.htm>; Lu F 1995). A 30-pound child who ate 3 ounces of cranberries contaminated with 0.45 ppm amitrole (three times the criterion for contamination) would have received six times the ADI that day. In 1959, cranberry juice was not yet popular, and cranberries were generally eaten only around the holidays.

Did Lawless regard this threat as overemphasized? Yes.

Media coverage: Hyped. "The use of the news conference format [before Thanksgiving] to announce the cautionary 'non-ban' gained a tremendous news coverage that successfully dramatized for the public and Congress the import of the Delaney Clause" (Lawless 1977, 67).

Media score: 196.

Was warning derived? Yes, from the recently passed Delaney Clause, DES.

CASE 3. DES IN LIVESTOCK

Nature of the warning: Initial warnings about DES concerned its use to fatten chickens, cattle, and sheep. In 1971, DES was reported to cause vaginal cancer in the daughters of women who had used it as a drug during pregnancy to prevent miscarriage.

First public warning: *New York Times,* January 29, 1956, p. E9, "Hormone and Cancer," reports a warning by four physicians at an FDA symposium against feeding DES to livestock because of the risk of cancer in humans who consume the meat.

Type of news source: Scientific, not peer reviewed (Dr. William Smith and three other physicians).

Did source have identifiable preexisting bias? Yes. Dr. Smith actively opposed any cancer-causing additives in food, even trace doses. He orchestrated the International Union Against Cancer, which reflected these views (see Marcus 1994).

Risk assessment: True (R). Laboratory animal studies linked DES to cancer in the early 1970s. In 1979, FDA prohibited the use of DES in U.S. food-producing animals.

Did Lawless regard this threat as overemphasized? Yes.

Media coverage: Hyped. On December 10, 1957, HEW Secretary Flemming made headlines with his announcement that the sale and use of DES for chickens—not beef or sheep—would be halted, and that the U.S. Department of Agriculture would buy and remove from the market $10 million worth of treated poultry. This came a month after Flemming's cranberry announcement (Lawless 1977, 74).

Media score: 152.

Was warning derived? Yes, from hearings on the Delaney Clause (finally enacted in 1958). Delaney publicized Smith's warning (*New York Times,* 1957, 10).

CASE 4. CYCLAMATE

Nature of the warning: Cyclamate is an artificial sweetener used especially in soft drinks. Its metabolite, cyclohexylamine, was thought to be carcinogenic in rats. HEW Secretary Finch, citing the Delaney Clause, banned its use on October 18, 1969.

First public warning: *New York Times,* December 14, 1968, p. 1, "Sugar Substitute Brings A Warning."

Type of news source: Government. "For the first time the cyclamate issue really came to the public's attention" through an FDA press release on December 13, 1968, restricting the use of cyclamate, affecting soft drinks. (Lawless 1977, 97). This followed a 1968 NAS report on artificial sweeteners, which said that totally unrestricted use was not warranted. Information also came from findings of the FDA's Dr. Legator on cyclohexylamine and Dr. Verrett on chick embryos.

Did source have identifiable preexisting bias? No.

Risk assessment: False (E). Extensive subsequent laboratory testing has consistently shown that cyclamate is not a carcinogen. It remains banned in the United States but is permitted in many other nations. (See Mazur and Jacobson 1999, reprinted in Appendix II, for details).

Did Lawless regard this threat as overemphasized? Yes.

Media coverage: Hyped. Dr. Jacqueline Verrett of FDA showed thalidomidelike birth defects in chick embryos on NBC evening television news, October 1, 1969.

Media score: 230.

Was warning derived? Yes, from thalidomide and the Delaney Clause. According to Lawless, "The media recognized the analogy between Dr. Verrett and Dr. Kelsey, heroine of the thalidomide tragedy" (Lawless 1977, 88).

CASE 5. MSG

Nature of the warning: Letters to the editor in the April 4, May 16, and July 11, 1968, editions of the *New England Journal of Medicine,* not all completely serious, described numbness and palpitation suffered by some patrons of Chinese restaurants and suggested that MSG, an additive common in Asian food, was a possible cause. Much greater concern was aroused in 1969 with Ralph Nader's warning that MSG in baby food was hazardous to infants.

First public warning: There were minor news reports on MSG in 1968. (Lawless's citation (1977, 97) of a front page story in the *Wall Street Journal* on July 11, 1968, is incorrect.) "Much greater public concern was aroused in 1969 when it was reported that MSG had caused brain damage when injected into newborn mice in large doses, because MSG was being widely used in human prepared baby foods" (Lawless 1977, 94). The first of these articles was "Nader Questions Safety of Baby Food Additives" (*New York Times,* 1969c).

Type of news source: Citizen. Ralph Nader warned in congressional testimony about MSG (and salt) in baby food. Nader possibly based this warning on J. Olney's "Brain Lesions, Obesity, and Other Disturbances in Mice Treated with Monosodium Glutamate" (1969).

Did source have identifiable preexisting bias? Yes. Nader opposed large corporations, including manufacturers of baby food.

Risk assessment: False (E, R). In 1987, WHO placed MSG in the safest category of food ingredients, along with items such as salt and vinegar. FDA also classifies MSG as "generally recognized as safe," which applies to most people when eaten at the customary level. Infants metabolize glutamate as efficiently as adults and do not display any special susceptibility to elevated oral intakes. Some people, especially those with asthma, might have transient symptoms if they consume very large quantities (FDA 1995).

Did Lawless regard this threat as overemphasized? No.

Media coverage: Hyped. Ralph Nader testified to Congress July 15, 1969, on MSG (and salt) in baby food. "On October 23, [1969] *The New York Times* gave front-page coverage to pleas by Dr. Olney and Dr. Jean Mayer (a Harvard researcher and President Nixon's advisor on nutrition) that MSG be removed from baby foods" (Lawless 1977, 99).

Media score: 92.

Was warning derived? Yes, from cyclamate.

CASE 6. BOTULISM

Nature of the warning: The death of a New York banker and the paralysis of his wife from botulism 28 hours after eating Bon Vivant soup raised fears that other cans of soup were poisoned.

First public warning: *New York Times,* July 2, 1971, p. 35, "Botulism Death in Westchester Brings Hunt for Soup." The next day an article, "Vichyssoise Alert Is Issued by State," appeared on page 1.

Type of news source: Government. Dr. Hollis Ingraham, New York State health commissioner, and the FDA confirmed botulism in the can from which the couple ate.

Did source have identifiable preexisting bias? No.

Risk assessment: True (E). Botulism was confirmed in blood samples from the victims and in some soup cans. In August 1971, Campbell Soup Company reported that a few of its cans were contaminated with botulin.

Did Lawless regard this threat as overemphasized? No.

Media coverage: Hyped. This incident was widely publicized by the news media and created much alarm (Lawless 1977, 106).

Media score: 115.

Was warning derived? No.

CASE 7. FISH PROTEIN CONCENTRATE

Nature of the warning: FDA banned whole fish flour in the United States, not fish flour per se, because it included parts that might be impure. By 1966, FDA's only objection was that too much fluoride content in the flour might mottle teeth.

First public warning: *New York Times,* January 25, 1962, p. 28, "Whole Fish Flour Is Barred for U.S."

Type of news source: Government (FDA).

Did source have identifiable preexisting bias? No.

Risk assessment: False (R). As of 2001, there are no FDA restrictions on fish protein powders.

Did Lawless regard this threat as overemphasized? No.

Media coverage: Routine.

Media score: 79.

Was warning derived? No.

CASE 8. FLUORIDATION

Nature of the warning: By 1950 a national network of opponents warned that the practice of adding fluoride to community drinking water to reduce the incidence of dental cavities was hazardous.

First public warning: There is no obvious "first warning" in the national news media. Local protest was well established by 1951 when letters to the editor in the *New York Times* began warning of fluoridation. The *Times* finally carried a news article, "Fight over Fluoride," on March 10, 1952.

Type of news source: Citizen. Individuals in the New York area wrote the first letters to the *New York Times*. Controversy in Seattle was the focus of the first *Times* article on March 10.

Did source have identifiable preexisting bias? Yes. Warnings were embedded in opposition to socialized medicine and big government.

Risk assessment: False (E, R). The most recent empirical evaluation by an authoritative scientific body is "Health Effects of Ingested Fluoride" (NAS 1993). The U.S. Public Health Service (PHS) recommends fluoridation.

Did Lawless regard this threat as overemphasized? Yes.

Media coverage: Hyped. Opponents typically used local news media as a platform for opposing fluoridation (Crain, Katz, and Rosenthal 1969).

Media score: 968.

Was warning derived? Yes, according to Mazur (2001), reprinted in Appendix II), reflecting concerns about socialism and communism in the United States. Lawless does not mention specific derivatives but notes, "A national network of antifluoridationists had come into being about 1950 and the issue was soon rephrased in terms of the rhetoric of the era" (Lawless 1977, 122).

CASE 9. SALK POLIO VACCINE

Nature of the warning: Amid the enthusiasm following the successful field test in 1954 of Dr. Jonas Salk's polio vaccine, which was based on killed virus, the PHS reported cases of polio in children who had recently received shots from a batch made by Cutter Laboratories.

First public warning: "The Cutter cases broke into the newspapers on April 27, 1955 .... As the news came out, public confidence in the Salk vaccine changed into doubt and then into fear" (Lawless 1977, 135). The first *New York Times* story appeared on April 28, 1955, p. 1, "One Firm's Vaccine Barred; 6 Polio Cases Are Studied."

Type of news source: Government. U.S. Surgeon General Leonard Scheele reported polio cases from California and Chicago associated with the Cutter vaccine.

Did source have identifiable preexisting bias? No.

Risk assessment: True (E). Two vaccine pools manufactured by Cutter Laboratories were implicated as the cause of polio in children receiving shots. The flawed manufacturing process allowed live virus to remain in the vaccine. Epidemiological investigation connected polio cases to the vaccine. For details, see Nathanson and Langmuir (1963).

Did Lawless regard this threat as overemphasized? No.

Media coverage: Hyped. Events surrounding the introduction and hurried distribution of the Salk vaccine received extraordinary press attention. The results of the 1954 field test were announced simultaneously to journalists, scientists, and the public on national television (see Wilson 1963).

Media score: 326.

Was warning derived? No.

CASE 10. THALIDOMIDE

Nature of the warning: This sleeping pill was approved for general use in Europe but not in the United States. "The thalidomide story was headline news in Europe during the winter [1961–1962] but received little attention in the United States. In February 1962, *Time* magazine described the European epidemic of phocomelia." Lawless also noted that pills were available in the United States "under heavy restrictions" (Lawless 1977, 144). This became a sensational news story in July after all tablets had been quietly recalled.

First public warning: *Time* magazine, February 23, 1962, p. 86, "Sleeping Pill Nightmare."

Type of news source: Scientific, peer reviewed. Dr. A. Speirs, "Thalidomide and congenital abnormalities," *Lancet*, February 10, 1962, 303–305; also Dr. W. Lenz of Germany.

Did source have identifiable preexisting bias? No.

Risk assessment: True (E). Thalidomide causes severe birth defects when taken during pregnancy. Today it is used to treat a complication of leprosy (www.fda.gov/cder/news/thalinfo/thalidomide.htm).

Did Lawless regard this threat as overemphasized? No.

Media coverage: Hyped. The *Washington Post* (1) and *Times Herald* of July 15, 1962, devoted prominent attention to the FDA's Dr. Frances Kelsey and her role in keeping thalidomide off the U.S. market (cited in Perlmutter 1962). The media also widely followed American Sherry Finkbine's battle to have a legal abortion after taking thalidomide in the early stage of her pregnancy (*Life* August 10, 1962, 24–36).

Media score: 261.

Was warning derived? No.

CASE 11. HEXACHLOROPHENE

Nature of the warning: In the late 1960s, hexachlorophene (HCP) was widely used as "pHisoHex" (3% HCP) for bathing infants, and as an antibacterial in soaps, deodorants, and cosmetics. An EPA study showed that HCP caused brain damage in animals.

First public warning: *New York Times*, August 22, 1971, p. 28, "Detergent Solution Used in Baby Baths Viewed as a Peril," reported that HCP can be absorbed through skin in quantities approaching toxic levels in rats.

Type of news source: Scientific, peer reviewed. The *Times* story was based on a recent article in the British medical journal *Lancet*, "Dermal Absorption of Hexachlorophane in Infants," by Drs. A. Curley, R. Hawk,

R. Kimbrough, G. Nathenson, and L. Finberg (August 7, 1971, 296–297). The first three authors worked at the FDA's Toxicology Branch (which in 1971 was shifted to the new EPA) and began their evaluation of HCP because of a manufacturer's application to FDA to use the chemical as a fungicide. However, the *Times* cites *Lancet* as its source and does not mention the FDA. No government agency had taken any warning action at that time. Indeed, the FDA was later criticized for ignoring HCP (Wade 1971). Furthermore, coauthors Nathenson and Finberg were nongovernmental scientists who worked at Albert Einstein College of Medicine. Since this warning reached the media through conventional scientific rather than government channels, its source is coded "scientific, peer reviewed."

Did source have identifiable preexisting bias? No.

Risk assessment: True (R). FDA limits the amount of HCP in cosmetics and over-the-counter drugs to 0.1% (vm.cfsan.fda.gov/~lrd/cf250250.html).

Did Lawless regard this threat as overemphasized? No.

Media coverage: Routine.

Media score: 48.

Was warning derived? Yes, from cyclamate, MSG, and mercury in tuna.

CASE 12. DMSO

Nature of the warning: An inexpensive industrial chemical, dimethyl sulfoxide (DMSO), received widespread attention as a miracle drug. However, FDA limited clinical tests, at first on nonspecific grounds, later halting clinical trials altogether for fear that DMSO might cause eye damage. Critics charged the FDA with suppressing a cheap and beneficial drug.

First public warning: *New York Times*, April 3, 1965, p. 28, ran an editorial, "DMSO—Promise and Danger," urging caution in the use of DMSO after the FDA's Dr. Francis Kelsey, heroine of the thalidomide episode, restricted human trials of the "wonder drug" to skin applications.

Type of news source: Probably government. The *New York Times* editorial does not cite a specific source but refers to the FDA's Francis Kelsey. It is likely that Kelsey or someone else in or close to the FDA was the source for the editorial supporting her action.

Did source have identifiable preexisting bias? Yes. Kelsey and others at FDA were very cautious after thalidomide. Professor Jack de la Torre of the University of New Mexico Medical School claims, "FDA had a sort of chip on its shoulder [in 1965] because it thought DMSO was some kind of snake oil medicine. There were people there who were openly biased against the compound even though they knew very little about it. With the new administration at the agency, it has changed a bit" (<www.dmso.org/articles/information/muir.htm>).

Risk assessment: False (E). The first warning urged caution because uncontrolled use of cheap (and not necessarily pure) DMSO was rapidly increasing. This warning was not based on specific evidence of harm but on an analogy to thalidomide, which had attained widespread use in Europe before its harmful effects were discovered. Seven months afterward, three laboratories noted alterations of the refractive index in the eyes of animals given large doses over a long period. No similar changes in human eyes had been reported, nor was it clear that these were adverse effects. Subsequent studies showed little indication of ocular or other harm at clinical doses (Leake 1966). FDA reinstituted full clinical trials in 1968. The National Academy of Sciences concluded in 1973 that DMSO effects on animal lenses were not appearing in humans at the doses tested but did not discount the possibility of a long-term hazard to humans (NAS 1973, 19). The *Physicians' Desk Reference* (1998), sent to me in June 1999 by the FDA in response to my inquiry about the status of DMSO, includes a "precaution" for Rimso-50, the sole FDA-approved version of DMSO, recommending that users receive full eye exams before and during treatment, but does not mention eye problems under "adverse effects."

Did Lawless regard this threat as overemphasized? Yes.

Media coverage: Routine.

Media score: 28.

Was warning derived? Yes, from thalidomide.

CASE 13. SHOE FLUOROSCOPES

Nature of the warning: Fluoroscopic shoe-fitting machines were widely used in shoe stores, exposing salespeople, children, and parents to x rays.

First public warning: *Time,* September 9, 1949, p. 67, "Little Feet, Be Careful."

Type of news source: Scientific, peer reviewed. Two research articles in the *New England Journal of Medicine,* September 1949, were picked up by *Time.*

Did source have identifiable preexisting bias? No.

Risk assessment: True (R). Shoe fluoroscopes are now banned as a needless exposure to ionizing radiation. (See Mazur 2000, reprinted in Appendix II.)

Did Lawless regard this threat as overemphasized? No.

Media coverage: Routine.

Media score: 28.

Was warning derived? No.

CASE 14. MEDICAL X RAYS

Nature of the warning: A 1956 report by NAS stating that the threat to human health from fallout was less than that from dental and medical x rays was widely covered in the news.

First public warning: *New York Times,* June 13, 1956, p. 1, "Scientists Term Radiation a Peril to Future of Man."

Type of news source: Scientific, peer reviewed. NAS report, *Biological Effects of Atomic Radiation,* June 13, 1956.

Did source have identifiable preexisting bias? No.

Risk assessment: True (R). (See Mazur 2000, reprinted in Appendix II).

Did Lawless regard this threat as overemphasized? Yes.

Media coverage: Hyped. "The [NAS] press conference and report created an immediate sensation in the news media, which was supplemented by numerous comments of agreement and disagreement by various authorities" (Lawless 1977, 192–193).

Media score: 173.

Was warning derived? Yes, from fallout controversy.

CASE 15. RADIATION FROM DEFECTIVE TELEVISIONS

Nature of the warning: On May 18, 1967, General Electric Corporation announced a recall of some 100,000 of its large-screen color television sets because they were found to emit low levels of x rays toward the floor.

First public warning: *New York Times,* May 19, 1967, p. 17, "General Electric Will Modify 90,000 Large Color Television Sets Against Possible X-Radiation Leaks."

Type of news source: General Electric. On its face, this is a rare example of a corporation warning the public about one of its own products. In fact, General Electric was forced to go public. In September 1966, GE began selling a new line of color televisions. In November, a GE inspector measured a narrow beam of x rays coming from the bottom of some sets, which routine quality-control checks of the front, top, and sides had failed to detect. Over the next few months the company, in consultation with government agencies, attempted to quietly correct the problem. On May 11, 1967, the National Center for Radiological Health (NCRH) of the PHS suggested that General Electric announce a recall program. On May 17, General Electric told government officials of new tests showing unacceptably high levels of unshielded x ray emissions. The company again indicated no plans for a public announcement, nor did the NCRH apparently intend to release a public statement immediately. The story was leaked to the press, and when the *New York Times* inquired about the matter on May 18, the NCRH and General Electric both made public statements, which were published jointly (Lawless 1977, 203).

Did source have identifiable preexisting bias? No.

Risk assessment: True (R). Leakage far exceeded today's FDA standard for television receivers of 0.5 milliroentgens per hour at a distance of 5 centimeters.

Did Lawless regard this threat as overemphasized? Yes.

Media coverage: Routine.

Media score: 28.

Was warning derived? Yes, from fallout controversy and warning about medical x rays.

CASE 16. SMOG IN DONORA

Nature of the warning: On Saturday, October 30, 1948, heavy smog settled over grimy, industrial Donora, Pennsylvania, 30 miles south of Pittsburgh, and lasted for several days. Twenty people died.

First public warning: "News of the Donora incident appeared on the front page of the *New York Times* for two days [October 31, 1948 (AP), and November 1, 1948] and received widespread public attention" (Lawless 1977, 221).

Type of news source: Government. Health officials involved in the emergency response were important sources, but the *Times* (and local newspapers) also used other medical or on-site sources for this breaking news story.

Did source have identifiable preexisting bias? No.

Risk assessment: True (E, R). No doubt the smog was the proximate cause of death, although it was not determined what particular component was most to blame (Roueche 1953). Air-quality standards now prohibit such high levels of air pollution.

Did Lawless regard this threat as overemphasized? No.

Media coverage: Routine.

Media score: 37.

Was warning derived? No. (I discount Lawless's connection of the event to a deadly smog episode in Belgium, in 1930, as too distant in time and place.)

CASE 17. MERCURY POLLUTION FROM INDUSTRY

Nature of the warning: In 1970, high mercury levels in Canada's Lake St. Claire and in Lake Erie were traced to contamination from factories in Canada and Michigan. The scope of pollution soon widened to other plants.

First public warning: *New York Times,* April 3, 1970, p. 43, "Lake's Fish Searched for Poison Mercury."

Type of news source: Government. FDA is the proximate source, warning that fish from Lake St. Clair should not be eaten and announcing that the search for mercury contamination would be extended to some of the Great Lakes. Although not mentioned by the *Times,* the ultimate source was Norvald Fimreite, a graduate student in Canada, who discovered pickerel in Lake St. Clair that averaged mercury levels of 1.4 ppm. U.S. and Canadian governments found that pollution from plants owned by Dow and Wyandotte plants contained mercury (Lawless 1977, 253).

Did source have identifiable preexisting bias? No.

Risk assessment: True (R). The EPA, in *Water Quality Criterion for the Protection of Human Health: Methylmercury* (2001), recommends that methylmercury in fish should not exceed 0.3 ppm. This is a health standard for water quality, not for the consumption of fish. EPA regards the concentration of methylmercury in fish as an indicator of water quality. (FDA permits methylmercury in fish up to 1.0 ppm.)

Did Lawless regard this threat as overemphasized? No.

Media coverage: Hyped. "[T]he tragedy of the Huckleby family in New Mexico became a national sensation when it was described on the February 17, 1970, Huntley-Brinkley network television newscast and in the press. Huckleby had butchered for his family's use one of the hogs that had been fed mercury-treated seed grain. The tragic result was that the Huckleby children all suffered severe brain damage with various degrees of permanent crippling, blindness and other disabilities" (Lawless 1977, 253). By summer, the issue had been taken up by government and media as "a major environmental scandal" (*New York Times* editorial, 1970d). *New York Times* ran page-one stories about the Hucklebys on August 10, 1970 (*New York Times* 1970f), September 11, 1970 (*New York Times* 1970g), and October 30, 1970 (*New York Times* 1970h).

Media score: 95.

Was warning derived? Yes, from DDT and, according to Mazur, from concern over the "death" of Lake Erie from eutrophication.

CASE 18. MERCURY IN TUNA

Nature of the warning: On December 15, 1970, FDA announced a recall of canned tuna because of mercury pollution, and later recalled most swordfish. (These fish are at the top of the food chain and biomagnify mercury.) Tests on museum specimens of tuna and swordfish indicated that these fish have always had mercury levels near the FDA safety limit.

First public warning: *New York Times,* December 16, 1970, p. 1, "Mercury in Tuna Leads to Recall." FDA announced a recall of almost 1 million cans of tuna, estimating that 23% of the catch was contaminated with excessive mercury (greater than 0.5 ppm).

Type of news source: Government. FDA Commissioner Charles Edwards Jr. announced the recall and is the source for the December 16, 1970, warning. FDA action was spurred by Professor Bruce McDuffie, a chemist at SUNY-Binghamton, who had measured 0.75 ppm of mercury in a can of tuna from his pantry, well above the 0.5-ppm limit set by FDA in 1969. Lawless released this information to the Binghamton newspaper. FDA called him and confirmed high levels of mercury in his tuna, eventually leading to its recall and warning.

Did source have identifiable preexisting bias? No.

Risk assessment: False (R). The percentage of tuna with greater than 0.5 ppm mercury turned out to be much lower than the 23% first thought. In 1985, FDA changed its regulatory basis from total mercury in fish to methylmercury, the most hazardous form, placing the action level at 1 ppm methylmercury in the edible portion of fish. As of January 18, 2001, FDA advises pregnant women against eating shark, swordfish, king mackerel, or tilefish, which could contain enough mercury to hurt a fetus's developing brain. However, the agency offers no such advice regarding tuna (*New York Times,* 2001). In December 2003, a scientific advisory panel told the FDA that it should provide clearer advice to pregnant women and young children on the risks from mercury in their diet, particularly specifying what types of tuna have more or less mercury. This raises the prospect that the FDA may in the future advise pregnant women and children to limit their consumption of certain types of tuna.

Did Lawless regard this threat as overemphasized? Yes.

Media coverage: Hyped. "The FDA's announcement (of December 15, 1970) caused a sensation during late December, reinforced by the FDA's recall on Christmas Eve of nearly every brand of frozen swordfish." (Lawless 1977, 263).

Media score: 87.

Was warning derived? Yes, from warnings about mercury pollution from industry, and from the Huckleby family's mercury poisoning (see Case 17, mercury from industry).

CASE 19. DDT

Nature of the warning: In 1949, Dr. Morton Biskind claimed that he had found a link between DDT and a "virus X" in humans (Lawless 1977, 273). In 1962, Rachel Carson associated chronic pesticide exposure with cancer and fertility problems in humans.

First public warning: Carson's *Silent Spring* (1962), serialized in three issues of *The New Yorker* and published as a book, was the first serious public alarm about DDT as a threat to human health.

Type of source: Citizen, Rachel Carson.

Did source have identifiable preexisting bias? Yes, Carson wrote her book as a polemic against pesticides.

Risk assessment: True (R). DDT is banned in the United States except for public health emergencies (www.atsdr.cdc.gov/tfacts35.html). EPA regards DDT as a probable human carcinogen (www.epa.gov/iris/subst/ 0147.htm). IARC concludes that there is inadequate evidence in humans for the carcinogenicity of DDT but sufficient evidence in experimental animals (193.51.164.11/htdocs/Monographs/Vol53/04-DDT.htm).

Did Lawless regard this threat as overemphasized? No.

Media coverage: Hyped. Media attention to *Silent Spring* was spurred by news that Dr. Frances Kelsey had prevented approval of thalidomide for U.S. sale, and by the interest of President John Kennedy. As Carson was dying from cancer, her book became a major focus of public controversy (see Chapter 3).

Media score: 1236.

Was warning derived? Yes, from cranberries, thalidomide, and, according to Mazur, from the weapons fallout controversy.

CASE 20. ASBESTOS

Nature of the warning: Airborne asbestos particles were linked to a rare cancer, mesothelioma, and possibly asbestosis and lung cancer.

First public warning: *New York Times*, May 28, 1970, p. 27, "City Acts to Curb Asbestos Spray," reporting that New York City was holding hearings on proposed regulations to control spraying of asbestos and detailing asbestos's link to cancer.

Type of news source: Scientific, peer reviewed. Dr. Irving Selikoff, Mt. Sinai School of Medicine, had published widely on the risk of asbestos.

Did source have identifiable preexisting bias? No.

Risk assessment: True (E, R). Asbestos is a human carcinogen (<www.epa.gov/iris/subst/0371.htm>). EPA regulates asbestos in air.

Did Lawless regard this threat as overemphasized? No.

Media coverage: Routine.

Media score: 72.

Was warning derived? Yes, from rising concerns about cancer and industrial pollutants, including mercury.

CASE 21. TACONITE POLLUTION

Nature of the warning: Reserve Mining Company was polluting Lake Superior with taconite (iron ore) tailings, possibly harming human health.

First public warning: *New York Times*, May 11, 1969, p. 65, "U.S. Report Cites Ore Plant Pollution," reported that Reserve Mining was dumping taconite in Lake Superior, and that the waste was contaminating the water supply of Duluth and other cities.

Type of news source: Government. Hearings of the Federal Water Pollution Control Administration, convened by Secretary of Interior Stewart in Duluth, Minnesota, May 1969, raised concerns about a health threat to drinking water from the taconite tailings.

Did source have identifiable preexisting bias? Yes. Local controversy over visual pollution of Lake Superior by Reserve's tailings had occurred well before the health warning.

Risk assessment: False (R). EPA does not regulate taconite in drinking water. EPA recommends a drinking water standard for iron of 0.3 milligrams per liter based on aesthetic effects, but does not require water systems to meet this limit (personal communication from EPA Safe Drinking Water Hotline, January 23, 2001).[1]

Did Lawless regard this threat as overemphasized? No.

Media coverage: Routine.

Media score: 9.

Was warning derived? Yes, from "death" of Lake Erie.

CASE 22. ENZYME DETERGENTS

Nature of the warning: Newly introduced enzyme detergents, touted for superior stain-removing power, were alleged to be a serious health hazard that caused dermatitis and asthma among consumers.

First public warning: *Time* magazine, February 16, 1970, p. 86, "Enzymes in Hot Water."

Type of news source: Apparently citizen. Lawless associates Ralph Nader with the warning (1977, 314). *Time* does not cite a specific source but reports the Federal Trade Commission's proposed study of health hazards from enzyme detergents. This apparently was a consequence of Nader's request that the FTC either ban the sale of enzyme detergents or label them as "serious health hazards" (*Chemical & Engineering News*, June 22, 1970, 14). This concern might have had some scientific basis. An article in the British medical journal *Lancet* (June 1969) reported that detergent plant workers, exposed to far higher levels than ordinary consumers, suffered asthmatic and skin conditions.

Did source have identifiable preexisting bias? Yes. Consumer advocate Ralph Nader is a frequent critic of large corporations.

Risk assessment: False (E, R). A National Academy of Sciences report of November 19, 1971, found no evidence of a major health hazard to consumers from enzyme detergents. "The evidence indicates that the average enzyme-detergent laundry product in normal use by consumers has not produced more irritation of the skin than have similar products that contain no enzymes." (See also "Study Detects No Peril in Enzyme Detergents," *New York Times*, November 19, 1971, 53.) Enzyme detergents are widely used today and not regulated by FDA or EPA.

Did Lawless regard this threat as overemphasized? Yes.

Media coverage: Hyped. Arthur Godfrey, a popular radio and television personality who advertised Colgate products, made news by publicly objecting to the company's new enzyme detergent as a water pollutant, thus spurring interest in Nader's health warning.

Media score: 48.

Was warning derived? Yes, from concerns about water pollution and other laundry detergents and, according to Mazur, from the "death" of Lake Erie.

CASE 23. NTA IN DETERGENTS

Nature of the warning: Introduced into detergents in 1970 as a replacement for phosphates, NTA was quickly suspected as a risk to human health.

First public warning: *New York Times*, November 13, 1970, p. 38, "Delay on Detergent Chemical Is Urged," reported Dr. Samuel Epstein's testimony to Congress that NTA might be a health hazard and urged more research before it was widely adopted.

Type of news source: Scientific, not peer reviewed. Dr. Samuel Epstein, Chief, Laboratories of Environmental Toxicology and Carcinogenesis, Children's Cancer Research Foundation, Boston, Massachusetts, had written an article warning on NTA for *Environment* magazine (1970) and later published a scholarly review of NTA in *International Journal of Environmental Studies* (1972).

Did source have identifiable preexisting bias? No. Epstein would later achieve a reputation as an extremist on the hazards of trace synthetic carcinogens (see Epstein 1998), but I have no evidence of prior bias regarding NTA or its producers.

Risk assessment: False (E, R). NTA is reasonably considered a human carcinogen based on animal evidence, but it biodegrades, so little or none is detected in drinking water. In 1980, EPA stated that it would not take regulatory action against the use of NTA in laundry detergent. NTA is currently used in two states where phosphates are banned and is widely used in Canada (<ehis.niehs.nih.gov/roc/>).

Did Lawless regard this threat as overemphasized? No.

Media coverage: Routine.

Media score: 44.

Was warning derived? Yes, from Lake Erie and problems with other detergents.

CASE 24. PLUTONIUM AT ROCKY FLATS

Nature of the warning: The Atomic Energy Commission's Rocky Flats plant near Denver, run by Dow Chemical Company, produced plutonium for nuclear weapons. On May 11, 1969, a major fire occurred at the plant. "Although apparently very little plutonium was actually lost as a result, the fire focused much attention on the operation of the ... plant and raised much controversy over whether it ... endangered the health and safety of nearby residents" (Lawless 1977, 337).

First public warning: Local coverage was well established when the *New York Times* first suggested a hazard in its article of June 27, 1969, p. 10, "Fire Cleanup Keeps Plutonium Plant Busy."

Type of news source: Citizen. Composed mostly of physicists, Colorado Committee for Environmental Information (CCEI) conducted an independent survey of radiation levels near the Rocky Flats facility. The *Times* describes CCEI as "a committee of Colorado environmentalists" (June 27, 1969, 10). It was part of the Scientists' Institute for Public Information network, started by Barry Commoner and Margaret Mead to provide scientific information relevant to policy problems. CCEI and its president, H. Metzger, played an advocacy role, participating actively in litigation and attempting to influence policy (Nelson 1970). I designated CCEI as "citizen" rather than a "scientific" source because of its political activism.

Did source have identifiable preexisting bias? Yes, CCEI participants had a "belief that government agencies (especially AEC) have been duplicitous"; two years earlier, CCEI had raised concerns about nerve gas stored at Rocky Mountain Arsenal (Nelson 1970, 1326).

Risk assessment: True (E, R). Weapons production was halted in 1989. The Rocky Flats site suffers has considerable radioactive contamination and is today a Superfund site. See <http://www.cdphe.state.co.us/rf/>.

Did Lawless regard this threat as overemphasized? No.

Media coverage: Routine.

Media score: 10.

Was warning derived? Yes, from Vietnam protest (Dow produced napalm) and the nuclear power controversy. CCEI, initially activated by the Dugway sheep kill incident, cited a similar danger of toxic release at Rocky Mountain Arsenal (Nelson 1970, 1326).

CASE 25. RADIOACTIVE WASTE STORED IN KANSAS

Nature of the warning: The Atomic Energy Commission (AEC) proposed an underground salt formation near the town of Lyons, Kansas, as the site for a permanent storage facility for long-term radioactive waste from nuclear reactors. Opposition from elsewhere in the state emphasized the possibility of radioactive leakage from the site.

First public warning: *New York Times*, February 17, 1971, p. 27, "Kansas Geologists Oppose a Nuclear Waste Dump."

Type of news source: Scientific (although arguably a government agency; see discussion in Chapter 6). The *New York Times* article refers to a report evaluating the geology and hydrology of the site prepared by the Kansas Geological Survey, a state agency housed at the University of Kansas. This report was "made public" by Ronald Baxter, chair of the Kansas Sierra Club. I designated the source "scientific" rather than "citizen" because Baxter acted simply as a conduit. As inferred by the headline, the focus of the article is the survey's report.

Did source have identifiable preexisting bias? No.

Risk assessment: True (E). Old drill holes for extracting oil and gas provided entry points for water, which could have carried radioactive material out from the salt. AEC abandoned the site as a repository (AEC press release P–20, January 21, 1972).

Did Lawless regard this threat as overemphasized? No.

Media coverage: Routine.

Media score: 23.

Was warning derived? Yes, from fallout controversy (Lawless 1977, 518) and, according to Mazur, the nuclear power controversy.

CASE 26. NUCLEAR TEST ON AMCHITKA

Nature of the warning: As part of the ABM development program, the AEC announced that an underground test of a 5-megaton nuclear device was to be conducted on the Alaskan island of Amchitka. Critics warned of a possibly disastrous earthquake or tsunami resulting from the blast.

First public warning: *New York Times*, April 30, 1970, p. 16, "U.S. Plan for New Aleutian Blasts Meeting Growing Opposition."

Type of news source: Citizen. During hearings held by the Alaska legislature, critics favored a resolution to bar the test. These critics included conservationists and a representative of the Federation of American Scientists, an advocacy group concerned mostly with issues related to nuclear weapons.

Did source have identifiable preexisting bias? Yes. The Amchitka protest was one episode in the ongoing protest against ABM development, which in turn reflected opposition to the war in Vietnam and the Nixon administration.

Risk assessment: False (E, R). No earthquake or tsunami followed the blast (Lawless 1977, 355). U.S. policy continues to allow underground testing of nuclear devices.

Did Lawless regard this threat as overemphasized? Yes.

Media coverage: Hyped. Reports on Amchitka tied to protests over ABM, the war in Vietnam, and Nixon made the front page of the *New York Times* on October 28, 1971, October 29, 1971, and November 7, 1971. The November 7, 1971, article covered a large demonstration in New York City's Central Park. A page-sized advertisement in opposition to the test, bought by John Gofman and others, appeared in the *New York Times* on November 5, 1971, p. 23.

Media score: 247.

Was warning derived? Yes, from protests against the ABM, the Defense Department, the war in Vietnam, and environmental degradation.

CASE 27. POISON GAS RELEASED AT DUGWAY

Nature of the warning: More than 6,000 sheep mysteriously died near the U.S. Army's Dugway Proving Ground in Utah in March 1968. The incident raised suspicions that the cause was a release of poisonous gas from the facility. The Army first denied, then acknowledged the possibility of an accident. Citizens were concerned about the facility's potential danger to human life.

First public warning: Local concern and press coverage were established by the time the incident was first mentioned in the *New York Times*, May 21, 1969, p. 10. "The widely publicized incident [in local media] created much public concern and suspicions were immediately aroused that the sheep had been killed by some lethal substance originating at the Dugway test site" (Lawless 1977, 360).

Type of news source: Citizen: sheep ranchers and veterinarians (Lawless 1977, 360).

Did source have identifiable preexisting bias? No.

Risk assessment: True (E). For a review of evidence that an accidental release of nerve gas did indeed kill the animals, see *Science* 162 (December 27, 1968), p. 1460, "Nerve Gas: Dugway Accident Linked to Utah Sheep Kill."

Did Lawless regard this threat as overemphasized? No.

Media coverage: Routine.

Media score: 56.

Was warning derived? No.

CASE 28. NERVE GAS DISPOSAL

Nature of the warning: In 1970, the news reported Army plans to transport obsolete chemical weapons (GB-Sarin), encased in concrete for safety, by train to New Jersey for ocean dumping. Critics claimed the operation risked accidental dispersal of nerve gas.

First public warning: *New York Times,* July 30, 1970, p. 1, "Army Will Transport Nerve Gas Across South for Disposal in Sea."

Type of news source: Government. Congressman Cornelius Gallagher (D, New Jersey) told journalists about the Army's plans for disposal and its risks.

Did source have identifiable preexisting bias? Yes. Gallagher "had been critical of the shipment of chemical and biological weapons throughout the country" (*New York Times* 1970e, 1).

Risk assessment: True (R). The United States discontinued ocean dumping of chemical weapons in the 1970s. Since 1986, the Army disposes of chemical weapons at their storage site. U.S. law now bans further transportation. Old chemical munitions may surface at old sites; this possibility is treated on a case-by-case emergency basis, with transport usually by air (Personal communication from Gregory Mahall, spokesman for the U.S. Army program manager for chemical demilitarization, April 11, 2001).

Did Lawless regard this threat as overemphasized? Yes.

Media coverage: Hyped. "During the summer of 1970 the papers were filled with news of the disposal program" (Lawless 1977, 373).

Media score: 118.

Was warning derived? Yes, from Dugway sheep kill; controversy over a shipment of nerve gas from Okinawa back to the United States for disposal; discussion of a Geneva Protocol to ban chemical and biological weapons; and also, according to Mazur, and protest against the war in Vietnam.

CASE 29. ELF RADIATION AT PROJECT SANGUINE

Nature of the warning: Residents of Wisconsin learned in 1969 of the Navy's Sanguine Project, a huge underground radio antenna to be built in their state for the purpose of communicating with nuclear submarines. Critics warned of health problems from the antenna's ELF radiation.

First public warning: *New York Times*, October 14, 1969, p. 49, "Controversy and Antennas Grow in Wisconsin."

Type of news source: Government. Louis Hanson, home assistant to Senator Gaylord Nelson (D, Wisconsin), reflected Nelson's concerns.

Did source have identifiable preexisting bias? Yes. The *New York Times* reported opposition mostly within the Democratic Party and among conservationists. The Navy's initial briefing of local residents on Sanguine in early 1969 aroused opposition from conservationists and antiwar activists. Senator Nelson, a strong environmentalist, was "incensed" at the Navy's tactics in siting the project in his state (Lawless 1977, 384–385).

Risk assessment: False (E). Two Nuclear Regulatory Commission reports, *Possible Health Effects of Exposure to Residential Electric and Magnetic Fields* (1997), and *Research on Power-Frequency Fields Completed Under the Energy Policy Act of 1992* (1999), found no convincing evidence that ELF radiation threatens human health.

Did Lawless regard this threat as overemphasized? No.

Media coverage: Routine.

Media score: 21.

Was warning derived? Yes, from Vietnam war protest, distrust of military, rising environmental movement.

CASE 30. CHEMICAL MACE

Nature of the warning: Chemical Mace, a form of tear gas in a hand-held sprayer, was developed in 1965 for riot control and quickly adopted by police. Controversy soon grew over whether Mace caused permanent injury.

First public warning: Article in the *New Republic,* April 13, 1968, by Roger Rapoport (recent editor of the University of Michigan's *Daily*), "MACE in the Face," p. 14.; also *New York Times,* May 11, 1968, p. 43, "Safety of Chemical Mace Is Questioned by Many."

Type of news source: Citizen: civil rights leaders, people sprayed with Mace in Michigan, and a University of Michigan dermatologist who examined one victim.

Did source have identifiable preexisting bias? Yes. Criticism of Mace is associated with civil protests of the time and opposition to police riot tactics.

Risk assessment: False (E, R). Permanent injury and death have occurred from severe exposure to tear gas, but there is no apparent hazard from transient incapacitating doses (<www.epa.gov/iris/subst/0537.htm>). Today Mace is used by many police departments, but some jurisdictions in the United States restrict its over-the-counter sale as a weapon.

Did Lawless regard this threat as overemphasized? Yes.

Media coverage: Hyped. "The controversy over mace might have waned, however, had it not been for the tumultuous Democratic National Convention held at Chicago in August (1968) .... Mace was being used and its use added to the furor." Then there was a "widely publicized" Senate hearing (Lawless 1977, 399).

Media score: 67.

Was warning derived? Yes, according to Mazur, from the civil rights movement and protest against the war in Vietnam. "The controversy ... primarily arose because of the use of weapons with a strong warfare connotation against ... well-educated and articulate ... demonstrators, at a time of increasing concern for civil rights" (Lawless 1977, 401).

CASE 31. INJURIES ON SYNTHETIC TURF

Nature of the warning: In 1971 the news media reported controversy over whether more injuries occur when football is played on synthetic turf than when it is played on natural grass.

First public warning: *New York Times,* September 1, 1971, p. 39, "Football Injuries Are Linked to Synthetic Turf" (Altman 1971).

Type of news source: Scientific, not peer reviewed. Dr. James Garrick, orthopedic surgeon and physician for the University of Washington football team, conducted a study of high school football inquiries on turf and grass, finding a higher injury rate on turf.

Did source have identifiable preexisting bias? No.

Risk assessment: True (E). Football injuries are more frequent on synthetic turf than on grass. (See Mazur and Bretsch 1999, reprinted in Appendix II.)

Did Lawless regard this threat as overemphasized? No.

Media coverage: Routine.

Media score: 26.

Was warning derived? No.

Note

1. In 1973, four years after the initial public warning, Reserve Mining's effluent was found to contain amosite asbestos, which EPA does limit to 7 million fibers per liter (fibers smaller than 10 micrometers).

APPENDIX II

CASE STUDIES

Reprinted here are four articles published in the journal *Risk: Health, Safety & Environment*, all retrospectives of warnings in the Lawless sample:

1. Allan Mazur and Jennifer Bretsch, 1999, "Looking Back: Synthetic Turf and Football Injuries," 10 (winter): 1–6.
2. Allan Mazur and Kevin Jacobson, 1999, "Looking Back: Cyclamate," 11 (spring): 95–100.
3. Allan Mazur, 2000, "Looking Back: Unneeded X Rays," 11 (winter): 1–8.
4. Allan Mazur, 2001. "Looking Back at Fluoridation," 12 (spring): 59–65.

LOOKING BACK: SYNTHETIC TURF AND FOOTBALL INJURIES

Allan Mazur and Jennifer Bretsch

Introduction

The Astrodome in Houston, Texas, completed in 1965, was the first of the huge covered arenas for professional football and baseball. Its domed roof had plastic windows that allowed sunlight to reach the grass on the enclosed field. But baseball fielders hated to play day games in the Astrodome because fly balls became invisible against the roof's mosaic of sunlit windows and dark supporting beams. The players missed routine catches and risked being hit on the head by the ball. To correct this problem, the windows of the dome were painted black, providing a uniform background. But the paint shut out the sunlight, so the interior grass died (Lawless 1977; Levy et al. 1990).

To solve *that* problem, the Astrodome's field was covered in 1966 with a nylon rug-like "grass" that had been developed by the Monsanto Chemical Company as a low maintenance surface for inner-city playgrounds. Monsanto promptly renamed its product "AstroTurf." Although some players disliked it from the outset, there were obvious advantages in reduced maintenance costs and durability, and the ease of changing the field for different sports, so other professional, collegiate, and high school arenas followed suit. AstroTurf remains one of the most widely used synthetic surface on athletic fields.

In 1968, Monsanto released statistics suggesting that the use of artificial turf could reduce football injuries by 80%, a plausible claim considering that natural grass hides holes and other field irregularities that can trip an athlete. But detractors charged that synthetic turf, despite its smoothness, *heightened* the risk of injuries among high-impact football players. The controversy became public near the beginning of the 1971 football season when *The New York Times* reported the claim of Dr. James Garrick, an orthopedic surgeon and team physician for the University of Washington football team, that the increased friction of dry synthetic turf made these fields more dangerous than natural grass gridirons. Garrick had conducted a one-year study of 139 football injuries in 228 high school games (80 on AstroTurf and 148 on natural grass) showing 0.76 injuries per game on the synthetic surface compared with 0.52 injuries per game on grass. The artificial turf had nearly a 50% higher rate of injury for the same weather during the same period. Garrick thought that knee sprains, especially torn ligaments and cartilage, were more likely when players' feet "hung up on the artificial turf," and that players ran faster and hit people harder on a surface that offers more traction (Altman 1971).

Monsanto immediately challenged the statistical validity of Garrick's results. The debate instigated an inconclusive Congressional investigation of synthetic turf. By this time the artificial surface was being used in twelve National Football League (NFL) stadiums, and the Players' Association formally demanded a moratorium on installations until the question of injuries was settled.

Risk Assessment

A spate of studies during the 1970s compared injuries on artificial turf and natural grass, but they gave inconsistent results because of methodological weaknesses, being based on small numbers of injuries and often failing to adequately control for the amount of player time on each kind of surface. A 1988 review of 32 studies from this period, comparing football injuries on the two surfaces, concluded that minor injuries (e.g. abrasions or bruises) were more common on synthetic turf, but with regard to severe injuries the results were equivocal. The knee and ankle are the most common sites of serious football injury. Nearly half the studies found the rate of knee and ankle damage to be at least 10% higher on turf than grass, and about half the studies found injury rates virtually equal (i.e., less than 10% difference) on the two surfaces. By this time no one seriously entertained Monsanto's initial claim that AstroTurf *reduced* injuries (Nigg and Segesser 1988).

Some controversies over risk assessment seem endless, with "truth" characterized as lying at the bottom of a bottomless pit. In this case the truth was rising toward the surface because of increasing experience with synthetic turf, especially as the NFL continued to accumulate injury statistics on turf and grass. By 1988 an article in the *Journal of the American Medical Association* could report on game injuries—usually to the knee or lower leg—suffered by the New York Jets professional football team during the period 1968–85 when about two-thirds of the games were played on grass, the rest on artificial turf. Counting only injuries serious enough for a player to miss at least two consecutive games, the Jets suffered 0.70 injuries per game on turf, and 0.59 injuries per game on grass. Injuries serious enough for a player to miss *eight* consecutive games occurred 0.36 times per game on turf, but only 0.23 times per game on grass (Nicholas et al. 1988).

A 1990 review of research by Skovron and her associates, about football injuries on turf versus grass, selectively focused on the methodologically strongest studies and data bases from high schools, colleges, and the NFL, concluded that "play and practice on artificial turf are associated with an increase in risk of time-loss injuries to the lower extremities .... Increased injury risks for other parts of the body have not been consistently demonstrated" (Skovron et al. 1990).

With the focus of concern now sharpening on the lower leg, new research by Powell and Schootman examined knee injuries in NFL games over the period 1980–89, when 2,572 team-games were played on grass, and 2,604 were played on AstroTurf. There were 1,081 knee sprains during these games, involving 860 players. On AstroTurf there were 0.223 sprains per game, on grass there were 0.196 sprains per game (Powell and Schootman 1992). Thus, the sprain rate was 13% higher on AstroTurf than on grass. This is a reliable difference, based on many games, but one that is small in magnitude.

Significance Tests

Why was there so much confusion and disagreement among the earlier risk assessors? Leaving aside those studies that were simply sloppy or highly biased, the main source of ambiguity may be illustrated by the 1988 *JAMA* article, discussed above, which showed that major injuries—serious enough for a Jet player to miss eight consecutive games—occurred 0.36 times per game on turf, but only 0.23 times per game on grass, a risk for turf that is over 50% higher than for grass. Remarkably, the authors of that study concluded, "there was no difference in the rates of ... major injuries per game ... between games played on grass or artificial turf ..." (Nicholas et al. 1988, 939). Obviously there *is* a difference in the rates; the authors meant there was no "significant difference," using the conventional statistical criterion, $p = .05$.

"Significant difference" has a technical meaning in statistics that is perplexing to laypeople. It arises when we want to describe a *population*, say, all the voters of New York State, but we have data only on a smaller *sample* of voters, randomly drawn from that population, as a pollster might do in trying to forecast the result of an upcoming election. We have no interest in the sample per se except insofar as it tells us something about the larger population. If we had data on the full population, as we do on election night, there would be no point in examining a sample.

Suppose 55% of a sample of New York voters, interviewed on the eve of the election, say they will vote for Gore, and 45% favor Dole. Can we say that Gore will win New York State on Election Day? We cannot say that with certainty because any sample may give a misleading result. But significance tests allow statisticians to judge the *probability* that Gore will win the election in New York, given his majority in the sample result. (By convention, if the statistician figures the probability of a tie vote, or a Dole win, to be less than five-out-of-a-hundred—$p = .05$—he will predict a Gore win.) But—this is important—statisticians can use the sample result to calculate the probability of a Gore victory *only if the sample has been randomly drawn from the population.* If the sample is not truly random, or very nearly so, is it useless as the basis for a significance test.

The problem of comparing injuries on grass and turf is similar. If we have a total population of football games, there is no need—indeed, no point—for significance tests. We simply see whether injuries are more frequent on turf than grass; it's like looking at the Gore–Dole vote after the election is over. However, if we have only a *sample* of turf and grass games, randomly drawn from a larger population, then we must calculate with significance tests whether or not a turf–grass difference seen in the sample is likely to be true of the population.

The authors of the *JAMA* study regarded the 18 seasons during which the Jets played on both grass and turf as a sample of seasons. Over those 18 seasons, injuries serious enough to miss eight games occurred 0.36 ± 0.29 times per game on turf (mean $\pm$ standard deviation), and 0.23 ± 0.16 times per game on grass. Using a two-tailed test of significance, they calculated that the difference between turf and grass rates was just shy of a "significant difference" at the $p = .05$ level; they therefore concluded "there was no [significant] difference in the rates."

The size of the sample is important: the larger the sample, the more likely an observed difference will be "significant" at the $p = .05$ level. Eighteen seasons are a small sample. Had the same Jet injury rates come from a somewhat larger sample of 26 seasons, they would have been "significantly different" (t-test, $p = .05$). Obviously, early studies, based on limited experience with turf, were less likely to show "significant" differences than later studies based on larger samples.

But the sources of ambiguity run deeper. Statisticians have some freedom to choose assumptions for their significance tests. Had the *JAMA* authors chosen a one-tailed rather than two-tailed test (permissible under the hypothesis that injuries are more frequent on turf), then they would have found a "significant difference" even with 18 seasons of data. Or if they had used different units of analysis, not seasons but, say, games, then the far larger sample size would have produced a more "significant" result.

Are significance tests relevant anyway? Are Jet seasons from 1968 to 1985 a random sample of a larger population, or do they comprise a whole population? If they are a random sample, from what population are they randomly drawn? To say Jet seasons are a random sample of all NFL seasons would be no more defensible than saying that the New York State vote for Dole and Gore is a random sample of the national election. On the other hand, one might conceptualize Jet seasons between 1968 and 1985 as an acceptable sample drawn from some imaginary population of all conceivable Jet seasons, in which case a significance test would be justified. This is a difficult issue that perplexes even statisticians.

With increasing experience on turf, this issue has become moot. Analysts who reject the appropriateness of significance tests can focus on the reliably higher injury rates on turf. Analysts who believe that significance tests are appropriate now have enough data to show a "significant difference" between turf and grass. Thus, the Powell and Shootman analysis, noted above, based on over 5,000 team-games in the NFL, showed knee sprains to be 13% more frequent on turf than grass *and also* calculated this to be a "significant difference" (using team-game as the unit of analysis).

Conclusion

The football database, especially from the NFL, has grown sufficiently to reliably show that synthetic turf increases the risk of knee sprain, and perhaps other injuries, although the magnitude of increase remains in doubt and may be small (Rodeo et al. 1990). Obviously, one should not assume that injury rates for the NFL would necessarily be the same for college or high school players. The National Collegiate Athletic Association (NCAA) has accumulated a large database on injuries from college sports but has not released an adequate analysis of the relationship between playing surface and football injury rate. The crude injury rate for

NCAA football, combining practice and games, is six percent higher on artificial surfaces than on grass over the period 1988–96, but this comparison has little meaning since confounding factors are not controlled (Dick n.d.).

LOOKING BACK: CYCLAMATE

Allan Mazur and Kevin Jacobson

Introduction

Cyclamates are a class of synthetic non-caloric substitutes for sugar, not as sweet as saccharin but without its unpleasant aftertaste. First approved in 1950 for use by diabetics and severely obese people, the Food and Drug Administration (FDA) in 1958 reclassified cyclamates as acceptable food additives, based on their history—a rather short one—of apparently safe use. At that time the American image of feminine beauty was busty and hippy, but during the 1960s, as slender figures became the ideal, the market increased for diet foods, especially diet sodas, and cyclamate use grew rapidly until it was being consumed by three-quarters of the American population (Lawless 1977; Mazur 1986).

The 1960s were also a time of increasing public concern about chemical adulteration of our food and water. The battle against *microbe* adulteration had largely been won in the early decades of the century, and scarcity was the major food concern during the Depression and World War II. In the bountiful 1950s we were warned to vary our diet so that we would get all necessary nutrients, as when television's popular science teacher, "Mr. Wizard," advised us incessantly to start each day with a "better breakfast" of *fruitcerealmilkbreadandbutterwitheggsorbreakfastmeatforvariety*. No concern here with fat or cholesterol, much less chemical additives or pollutants. We bought DDT "bugbombs" to destroy insects in our homes and gardens, while trucks and airplanes sprayed clouds of the inexpensive insecticide over fields and neighborhoods, often using far more than recommended, carelessly engulfing animals and people. Most of us didn't care then.

By the late 1950s, Americans on the political left were complaining of the hazard of trace doses of radioactivity from the atmospheric testing of atomic bombs. At nearly the same time, but on the political right, was a mass protest against adding fluoride—toxic in high doses—to public water supplies at one part per million to prevent dental cavities. Although there was little overlap in the memberships of these movements, their arguments against trace poisons were basically the same. In 1958 a clause introduced by U.S. Representative James Delaney, an antifluoridationist, was added to the Food, Drug, and Cosmetics Act, banning the addition to processed food of any chemical shown to cause cancer in an animal. These political movements paved the way for Rachel Carson's best-seller, *Silent Spring,* warning of trace pesticides. It appeared in 1962, just as the tranquilizer thalidomide was announced to cause birth defects when taken by pregnant women. Approved for use in Europe but not the United States, thalidomide had caused babies to be born with flipper-like stumps instead of arms and legs. Dr. Frances Kelsey of the FDA had single-handedly stood firm against great pressure and abuse in denying approval to thalidomide, thus saving America from the tragedy of armless and legless children. She provided the best press the FDA every had (Mazur 1996).

Cyclamate was among several food additives and drugs that became suspect during the 1960s (Turner 1970). The National Academy of Sciences, in periodic reviews for the FDA of cyclamate toxicity, found the sweetener generally safe but warned against uncontrolled distribution to the public because it has physiological effects including diarrhea, and the consequences of prolonged exposure were unknown. Also, doubts were raised that people using cyclamates actually lose weight. In 1968 FDA scientists led by Dr. Marvin Legator showed that cyclohexylamine, a metabolite of cyclamate that forms in the digestive tract, causes chromosome breakage in rats, suggesting the possibility of gene damage. Responding to these concerns in December 1968, the FDA reduced its recommended daily upper limit for cyclamate to 50 milligrams per kilogram of body weight, which meant that two cans of diet soda might exceed the limit for a 60-pound child.

In the meantime, another FDA scientist, Dr. Jacqueline Verrett, had been injecting cyclamate into chicken eggs, producing deformities in the embryos. These results were forcefully brought to the attention of FDA Commissioner Herbert Ley in April 1969 when one of Verrett's superiors carried deformed chicken embryos into Ley's office, but the commissioner took no action. On October 1, 1969, Dr. Verrett was interviewed on NBC television news, apparently with the consent of an FDA deputy commissioner. Setting off a wave of concern, she described her finding that cyclamate deforms chicken embryos, its thalidomide-like inference to human embryos fairly obvious. Verrett's appearance drew an immediate rebuttal from Commissioner Ley, saying "Cyclamates are safe within the present state of knowledge and scientific opinion available to me." Spokesmen for Abbott Laboratories, one of the largest cyclamate producers, also declared the sweetener safe and attributed the alarm to the sugar industry (Mintz 1969). Commissioner Ley's boss, Secretary of Health, Education and Welfare Robert Finch, was drawn into the controversy, criticizing in the press the FDA's handling of safety questions about cyclamate and indicating that reorganization of the agency's procedures and personnel was inevitable.

Two weeks later, Abbott Laboratories released a study it had commissioned showing that 8 out of 80 rats fed large amounts of a mixture of ten parts cyclamate to one part saccharin (the proportions most often used in diet sodas) developed bladder tumors, some of them cancerous. This was the basis for Secretary Finch's announcement, at a press conference on October 18, that indications of cancer clearly placed cyclamate in violation of the Delaney Clause. Therefore the FDA could no longer allow it to be used in nonprescription drinks and foods, and products with cyclamate must be recalled from stores. The following year, 1970, the FDA banned cyclamate from prescription products too.

Risk Assessment

It had not been cancer but the thalidomide-like birth defects in televised chick embryos that had spurred public concern over cyclamate in 1969. Today this is not an issue of scientific concern. Injection of a chemical into a chicken egg is devalued by authorities as a test for human birth defects because there is no parallel between the chicken embryo and a mammal's placenta. In any case, new feeding tests showed that cyclamate does not produce birth defects in mammals (Verrett et al. 1980; Bopp et al. 1986; Williams 1988).

The FDA's judgment that cyclamate may cause bladder cancer was based on one study of rats fed a combination of cyclamate and saccharin. (Tests where cyclamate materials were implanted into rodent bladders were generally discounted because implantation does not correspond with feeding as a route of administration.) By 1982, when Abbott Labs and a trade organization, the Calorie Control Council, petitioned anew for approval of the sweetener, some twenty new cancer bioassays had been conducted in which cyclamate was fed to a variety of animal species without showing any cancer-causing effect, even at high doses. Two attempts to replicate the original study linking cyclamate to bladder cancer failed to produce cancer in rats. The cause of the tumors in the original rats was never established, but conjectured causes include parasites or stones in the rats' bladders, or environmental contamination of some of their cages (Calorie Control Council 1982; Meister 1993).

In 1984 the FDA's own Cancer Assessment Committee concluded that cyclamate is not carcinogenic. The agency, under embarrassing pressure to reverse its ban, asked the prestigious National Academy of Sciences to evaluate the cancer causing potential of the sweetener. The academy concluded the next year, on the basis of epidemiological and animal studies, that neither cyclamate nor its metabolite cyclohexylamaine are carcinogenic. However the academy did not fully exonerate cyclamate, instead suggesting more investigation of hints that it could be a *co*-carcinogen, promoting the action of other cancer-causing substances that might be in the body (NAS 1985; the same conclusion had been reached by the FDA's Cancer Assessment Committee a year earlier).

While the cancer issue was moving toward resolution, the prospect of genetic damage remained ambiguous. Legator's 1968 finding, that the metabolite cyclohexylamine causes chromosome breakage in rats, could not be confirmed by other investigators. Many other tests were run using diverse methods; some supported the possibility of gene damage, some did not. This matter was complicated by great uncertainty at that time, still not fully resolved, about how to test for mutagenic effects. Scientists associated with Abbott or sponsored by the Calorie Control Council, evaluating the entire battery of tests, concluded that neither cyclamate nor its metabolite represent a significant mutagenic hazard (Bopp et al. 1986; Brusick et al. 1989). We are not aware of a serious challenge to their conclusion, but the issue may never be fully resolved.

Nothing can be proven 100% safe. By 1989 cyclamate seemed to have sufficiently passed muster. The front page of the *Washington Post* claimed the FDA was "widely expected to reapprove it, possibly this year. Once accused of causing everything from bladder cancer to birth defects, cyclamate is now widely thought to be harmless." Robert Scheuplein, acting director for the FDA's Center for Food Safety and Applied Nutrition, was quoted, "I have no reluctance in saying that with cyclamate we made a mistake." Referring to an earlier refusal by the FDA to lift the ban, back in 1980, Scheuplein said, "The matter was taken out of the hands of the scientists here and handled by attorneys" (Gladwell 1989).

Whether because of its attorneys or scientists, the FDA did not reapprove cyclamate in 1989 nor has it acted on the petition as of this writing, ten years later. New suspicions had arisen about cyclamate's metabolite. Male animals show testicular atrophy when fed cyclohexylamine at high doses. The petitioners responded that there would be no diminution of testicles if the acceptable daily dose of cyclamate were suitably low. Also, the FDA was concerned that the metabolite elevated blood pressure in test animals and humans, leading to a new round of research.

Conclusion

The diet soda industry, which had consumed roughly 60 percent of all cyclamates, quickly responded to the ban by substituting a saccharin-sugar mixture, and consumption of diet products continued upward. Industry criticized the Delaney Clause as being too restrictive, preventing any balancing of the benefits of an additive against its risks, but Delaney remains in effect. Citing Delaney, the FDA in 1977 proposed to ban saccharin too after animal research suggested that it caused bladder cancer, however Congress quickly set aside this ban, and today saccharin remains in use along with the newer sweetener, aspartame, and the newest one, acesulfame potassium.

Cyclamate does not cause cancer, or birth defects in mammals, as had once been feared. But as each suspected harmful effect was examined and ruled out, others surfaced: co-carcinogenicity, testicular atrophy, high blood pressure. From the perspective of the sweetener's defenders, this seems to be a carnival game of Whack-A-Mole; as soon as one mole is beaten back into his hole, another mole emerges from another hole, and then another. Cyclamate's critics, on the other side, may feel as if they are searching a pearl-rich bed of oysters, prying open one shell after another until the payoff eventually comes to light. There is no objective way for this kind of assessment to finally conclude that cyclamate is safe because there will always be more oysters to open and more moles to knock down. Safety can be affirmed only when the assessors lose interest in looking further, as may be happening today.

More than 50 nations, including most of Europe, approve cyclamate as a sweetener. The World Health Organization's Joint

Expert Committee on Food Additives has consistently determined cyclamate to be safe. Canada, in a twist from the American position, has more stringent restrictions on saccharin than on cyclamate, allowing both as table-top sweeteners but only cyclamate as a sweetening agent in drugs. Sweden, one of the nations retaining its ban on cyclamate, was told by the European Commission in 1999 that it was not entitled under European Union law to keep this restriction because the additive was not harmful to human health (Reuters 1999).

The petition for approval of cyclamate that Abbott Labs and the Calorie Control Council submitted to the FDA in 1982 is still pending. The FDA will not make public comments on pending petitions, so we do not know if it is under active consideration or not. Abbott Labs says it is no longer interested in the product, but a spokesperson for the Calorie Control Council tells us that a number of companies still want to use cyclamate in the U.S. In view of the considerable public suspicion about chemical sweeteners, probably including several now awaiting approval, the industry would no doubt enjoy seeing cyclamate's reputation restored, even if it never appears in U.S. markets.

LOOKING BACK: UNNEEDED X RAYS

Allan Mazur

Introduction

I recall as a child in the 1940s shopping for shoes with my mother. To check the fit, all of us—I, my mother, and the salesman—peered down into an x-ray fluoroscope while I wiggled my toes in the new shoes. A fluoroscope was also part of routine visits to the pediatrician: the doctor and my mother looked at the glowing screen showing the inside of my torso. This had no important diagnostic value for healthy-appearing children, but low level x rays were not then regarded as too costly for their entertainment value.

Thomas Edison set an assistant to work on the fluoroscope in 1896, only a year after Wilhelm Roentgen discovered x rays. The assistant's hair fell out after frequent exposure to the rays, and his hands become ulcerated and then cancerous; eventually the disease killed him. Because of such experiences, it was thought that radiation had to inflict ulceration or other gross damage in order to cause malignancy, and that as long as workers avoided dosages large enough to produce burns or other severe bodily changes, they and their patients were safe. With this mindset, early workers tested the functioning of their machines on their own hands as they began work each day, eventually accumulating massive doses.

By the 1920s, it was apparent from the large number of burns and other skin problems suffered by radiation workers, and some ensuing cancers, that safety standards were desirable. A British group led the way, recommending that x ray and radium workers limit their exposure by keeping distance and lead shielding between themselves and the radiation source, and that they work not more than 35 hours per week or have less than one month's holiday a year. Not until 1934 did the National Committee on Radiation Protection (NCRP) in the United States recommend a maximum permissible exposure level, fixing 0.1 rem (r) per day as a limit adequate to prevent overt skin damage and therefore, it was thought, more serious consequences (Taylor 1958; Mazur 1998).[1]

A few press warnings about shoe-fitting fluoroscopes followed the publication in 1949 of two articles in *The New England Journal of Medicine*. One reported that radiation exposures from these machines to customers and clerks often exceeded the maximum permissible dose. The other claimed that the most likely injuries from shoe fluoroscopy are malformation of growing feet, skin damage, and injury to the blood-forming tissues of store clerks, all dangers that can be controlled by proper regulation of the machines, as well as the posting of signs warning against too many exposures (Lawless 1977). There was no move to eliminate shoe fluoroscopy entirely.

In 1954 a Japanese fishing boat, the *Lucky Dragon*, was accidentally showered with fallout from an American hydrogen bomb test, precipitating first in Japan, then in the United States and Europe, a social movement aimed specifically at halting atmospheric testing. This anti-testing movement, more generally an expression of opposition to the arms race and nuclear confrontation with the Soviets, had an enormous effect on public perceptions of environmental radiation and its dangers and became a presidential election issue in 1956. Partly in response, and also anticipating the introduction of peaceful nuclear energy, the National Academy of Sciences (NAS) in 1956 reported that the U.S. population was being exposed to far less radiation from weapons-testing fallout than from either naturally-occurring sources such as cosmic rays or, surprisingly, from medical and dental x rays.

By this time it was known, mostly from experiments on fruit flies and mice, that sizable doses of x rays can produce genetic mutations in the descendants of irradiated individuals, and that the mutation rate increases with the dose of radiation. The NAS committee, dominated by geneticists, focused on this danger, asserting in its report more than had truly been demonstrated experimentally: that *any* amount of radiation to the gonads, however small, can cause mutations, and that "a little radiation to a lot of people is as harmful as a lot of radiation to a few, since the total number of mutant genes can be the same in the two cases." The committee estimated that the amount of radiation required to double the naturally-occurring mutation rate in humans was probably between 30 and 80 r. It recommended—without any clearly stated rationale—that average cumulative exposure of the population's gonads from man-made sources should not exceed 10 r from conception to age 30 (the mean age of reproduction). The committee estimated fallout exposure from weapons testing to be only one percent of this limit, while exposure from medical x rays and fluoroscopy—here was the kicker—was thirty times higher than from fallout! Dr. Warren Weaver, a principal of the report, warned at an NAS press conference, "We have been rather profligate about using x rays .... If anything is stupid from a genetic point of view, it is to use x rays to fit shoes on people." He also condemned obstetricians who x ray pregnant women merely to show them how "beautifully formed" is the skeleton of their baby (NAS 1956; *US News & World Report,* June 22, 1956, 64).[2]

These conclusions were widely reported, including a prominent story on the front page of the most important agenda-setter of the American press, *The New York Times* (June 13, 1956), which published the entire genetics portion of the report. Rather than lessening public worries about fallout, the effect of the report, and of media coverage that it triggered, was to extend these concerns to medical and dental x rays. Radiologists, in defense, blamed most over-exposure to x rays on faulty equipment or sloppy procedures by untrained technicians, saying the public was becoming unduly frightened of a valuable diagnostic procedure.

The anti-fallout movement ended abruptly in 1963 when President John Kennedy and General Secretary Nikita Khrushchev, frightened by their clash the previous year over Soviet missiles in Cuba, signed an agreement to halt the testing of nuclear weapons in the atmosphere, easing cold war tensions. Press and public concern with medical x rays, which had been borne upward by the fallout controversy, fell too, at least until 1967 when many color television sets were discovered to emit significant amounts of x rays, leading to hearings in the U.S. Congress that rejuvenated public concern (Lawless 1977, 197). In 1968 Congress made the Food and Drug Administration responsible for reducing unnecessary human exposure to man-made radiation from television and from medical and dental radiology.

By this time most states had banned the use of x rays on humans by anyone other than medical or dental personnel, and shoe fluoroscopy had fallen out of favor. The last functioning shoe fluoroscope, as far as I know, was discovered in 1981 in a West Virginia department store and finally donated for display in a medical museum. Gratuitous x rays in medical practice, including fluoroscopy in routine pediatric and obstetric examinations, were also eliminated, especially as sonograms provided an innocuous replacement. When medical or dental radiation was warranted, there was a clear trend toward increased shielding, better focusing of the beam, and reduced exposures permitted by more sensitive film. Marginal uses of the ray were reconsidered, notably the curbside radiology vans sponsored by tuberculosis associations, in which the public had been urged since the 1940s to get a yearly chest x ray (Nelson 1971). (By 1970 tuberculosis had declined to an extent that the disease was rarely discovered through x rays, and in any case, the tuberculin skin test had become the preferred and less expensive method of screening.) The possibility was raised that mammography screening for breast cancer might cause more tumors than it discovered among younger women. By this time, too, irradiation of children's throats and heads was stopped for the treatment of minor problems like enlarged tonsils, ringworm or acne, as it had recently been discovered that these exposures caused thyroid cancer many years later (in my own case, 28 years later; Mazur 1981).

Risk Assessment

The 1956 NAS committee knew from emerging data on leukemia in Japanese atomic-bomb survivors, and from malignancies in women who had painted radium on luminous watch faces, that ionizing radiation at high doses could cause leukemia and bone cancer. But it showed little concern about cancer from the low doses that the average person received from fallout, x rays, and natural sources, which it estimated at less than 10 r over 30 years. Instead the committee's emphasis was on the harm of genetic mutations to future generations, which it regarded as the primary problem of low doses.

The committee's reasoning is clear enough. A mutation in an egg or sperm cell might be caused by a single hit of ionizing radiation to the gonads; that could suffice to break one chemical bond in a gene, altering its function in all subsequent replications. Therefore any degree of radiation, however small, carried the possibility of mutation. Cancers, on the other hand, did not develop until years or decades after a person's exposure to radiation, making it implausible that carcinogenesis is a single-stage process, fully implemented by one hit of radiation. It seemed more likely that some minimum level—a threshold—of radiation exposure is required before cancers are induced. If the threshold were above 10 r, as was widely assumed, cancers would not be caused by the low exposures of concern to the committee in 1956.

In 1958 Linus Pauling published *No More War*, his polemic against atomic war and weapons testing, a book as important as *Silent Spring* in launching public concern about environmental contamination.[3] Pauling, winner two years earlier of the Nobel Prize in Chemistry, argued that there was no threshold, that even the lowest exposures cause cancer, just as they cause genetic mutation. He was not the first scientist to make these claims but certainly gave them the greatest visibility. Pauling estimated that if a population's radiation exposure were increased by only one rem, that small addition would produce two additional cases of leukemia per million people per year, and about one-fifth as many new bone cancers. He calculated that worldwide fallout from the atmospheric testing of a single large nuclear weapon eventually leads to the deaths of 10,000 people by leukemia and bone cancer, and possibly 90,000 more by other diseases, plus genetic effects in babies. These were controversial claims.

As public fears about fallout lessened after the 1963 Atmospheric Test Ban Treaty, the scientific debate about cancer hazards from low-level radiation remained of interest only to professionals. But the respite was brief, and by 1970 the debate had returned to the newspapers atop another public issue: the growing nuclear power industry. Drs. John Gofman and Arthur Tamplin, two health physicists at the Lawrence Radiation

Laboratory in California, accepted the assumption that radiation has no threshold for cancer, that even the smallest exposures are carcinogenic. They asserted to public audiences, including Congress, that if the American population were exposed to radiation from nuclear power plants at the maximum level permitted by federal regulations, there would be, each year, an additional 32,000 cases of leukemia and other cancers. Hearing echoes of Pauling, the nuclear industry nearly overheated with denial (Mazur 1981).

Responding to the tumult, the NAS again convened a committee of experts to re-evaluate the hazards of low-level radiation in the light of new knowledge, and on the assumption that cooler heads working outside the limelight might reach a rational conclusion. By 1966 an excess of roughly 100 leukemia cases had been observed among 117,000 survivors of the Hiroshima and Nagasaki bombings. From these and studies of patients who had gotten cancer from medical irradiation, it was clear that the number of malignancies produced in a population is proportional to the cumulative dose of radiation to the whole body. However this linear relationship was confidently observed only at sizable exposures, above 50 r. If excess cancers were being produced at lower doses, they were too few in number to be confidently recognized. Nonetheless, the new NAS committee concluded—consistent with Pauling, Gofman and Tamplin—that "the only workable approach to numerical estimation" of low-exposure effects is to extrapolate the linear dose–response relationship, known for sizable exposures, down to zero dose. On this assumption, the committee calculated that if the American population were exposed to radiation at the maximum level permitted by federal regulations, there would be, each year, an additional 6,000 cancer *deaths*. The committee's 1972 report recommended that the maximum permissible exposure to the public be lowered (it was) and sharply admonished the medical profession to voluntarily limit its use of x rays (NAS 1972; Gillette 1972).[4]

NAS committees have continued to assess biological effects of x rays. By 1985, fortunately, no statistically significant excess of genetic defects had been detected among the children of A-bomb survivors, but hundreds of excess cancers had appeared among the survivors themselves. Clearly, solid cancers as well as leukemia increase in frequency with increasing dose, but this has been confidently observed only at exposures above 10 r. The most recent NAS committee to evaluate x-ray effects reported in 1990 that it did not know the degree to which doses below 10 r produce cancer, or if x rays at such low doses cause any cancer at all. Nonetheless, this committee, like the others since 1972, extrapolated the dose–response curve down to zero dose, assuming no threshold (NAS 1990).

Conclusion

In 1956 dental x rays and fluoroscopy might deliver several rems in a single exposure, whereas a single shot today should be less than one rem. The 1990 NAS committee estimated that the average American is exposed to .04 r per year from diagnostic x rays. How many cancer deaths does this cause in the American population?

The 1990 committee, assuming no threshold, estimated that if one million people of all ages continuously received 0.1 r of x rays annually, this would cause about 5,600 extra cancer deaths during their remaining lifetimes (or about half a percent more cancer deaths than would normally be expected). At low doses, the committee's risk estimate is directly proportional to dose, so a yearly exposure of 0.04 r—the average American's annual exposure to diagnostic x rays—would eventually cause 2,250 extra cancer deaths per million people, or a total of 604,800 cancer deaths among the 270 million Americans alive today. On the other hand, if there *is* a threshold of one rem or higher, a possibility the committee does not deny, then there are zero extra deaths from diagnostic x rays (NAS 1990, 7, 171–181; at low doses, the committee's preferred risk models are linear).

Ambiguity may always remain about the harm caused by very low doses of ionizing radiation. The reason is that doses below 10 r are expected, even in the worst case, to cause so few cancers that it would take a huge number of irradiated individuals—far more than 100,000 A-bomb survivors (or laboratory mice)—to detect with confidence any increase above the far larger number of normally occurring cancers. Responding to this gap in knowledge, and mindful of the possibility of a threshold, the Health Physics Society, a professional organization promoting radiation safety, recommended in 1996 against any quantitative estimation of health risks for lifetime doses below 10 r above background radiation.

Happily, despite continued disagreement about actual effects, there has evolved a consensus among responsible authorities that unneeded x rays, even at low exposures, cannot be justified in view of the *possibility* of adverse effects. No one today would favor shoe fluoroscopy or the irradiation of fetuses. Furthermore, it is widely held that when x rays are needed, exposures should be kept as low as is reasonably achievable (e.g., NCRP 1999).

Footnotes

1. Prior to World War II, the roentgen was the common measurement unit for x rays. Newer units, the rad and rem, were defined to incorporate other forms of ionizing radiation, and these too have been superceded, confronting the historian with confusing terminology. Fortunately, these units are essentially equivalent for x rays, so I have expressed them all in terms of rems. The NCRP has evolved, under a slightly extended name, into a quasi-governmental authority on radiation risks.

2. The NAS committee seems to have chosen 10 r as a limit because it is the lowest round number above the 4 r that an average person receives from naturally-occurring background radiation. In 1957 the NCRP lowered its maximum permissible dose for the average person to the level recommended by the NAS.

3. Rachel Carson's *Silent Spring* (1962) is better remembered today, but Pauling's book was as important at the time of its publication, contributing to his receipt of the Nobel Peace Prize in 1962. Carson adapted the imagery of radioactive fallout, famously picturing pesticides as white powder, falling like snow on the roofs of an American town, causing a mysterious affliction that kills first animals, then humans.

4. The NAS committee estimated the effect of radiation on leukemia to be half of Pauling's estimate, and it discounted radiation-induced thyroid cancers because they are not usually fatal.

LOOKING BACK AT FLUORIDATION

Allan Mazur

Introduction

Fluoridation was the first technology after the Second World War to arouse widespread public opposition, opening an era of modern politics marked by disputes between experts over factual matters of risk and efficacy.

In the 1930s it was noticed that people living where the drinking water naturally contained fluoride had teeth that were often discolored but also relatively free of cavities. Further work showed that the benefit of cavity prevention could be enjoyed with little discoloration if the concentration of fluoride was as low as one part per million (ppm). In 1945 the U.S. Public

Health Service (PHS) began experimentally adding fluoride at this concentration to the drinking water of a few cities, intending over the next ten years to compare their cavity rates to those of control cities. A group of Wisconsin dentists, enthusiastic over the low cavity rates reported during the first years of the study, urged that mass fluoridation be promoted immediately. The PHS first resisted, saying it would wait until the completion of the ten-year experiment, but in 1950 approved nationwide fluoridation. By 1951 the American Dental Association and the American Medical Association had added their endorsements, urging communities to fluoridate (McNeil 1985).

Almost immediately, politically conservative citizen groups in Wisconsin protested against adding a toxic chemical to their drinking water, arguing that fluoride (in higher doses) was a rat poison and that involuntary fluoridation amounted to mass medication, a step toward socialism. The movement spread across the United States, gaining strength from concerns the federal government was susceptible to communist influences, then to other nations, though my treatment is limited to the American case. When American communities voted in referenda whether or not to fluoridate, usually the measure lost (Crain et al. 1969).

There is an exaggerated stereotype of the antifluoridationist as a kook, a fanatic right-winger, as represented by the mad General Jack D. Ripper in Stanley Kubrick's film *Dr. Strangelove*. Few "neutral" commentators gave serious consideration to the arguments of the opponents because they had been successfully painted by establishment proponents as irrational extremists. Psychologists of the time called opposition to fluoridation an "anti-scientific attitude," and social scientists viewed referendum defeats as democracy gone astray (Mausner and Mausner 1955; American Dental Association 1965).

But among the opponents of fluoridation were respectable scientists, physicians, and others sensibly cautious about chronic toxic effects. From today's perspective, it was reckless of the PHS and other health organizations to promote mass fluoridation as early as they did. Fluoride is a known poison at high dosage, and data on humans then used to evaluate the health risk of adding a small amount to drinking water were limited to crude comparisons of vital statistics among selected communities with varying levels of naturally occurring fluoride, and to pediatric examinations on children in one of the experimental cities exposed to fluoride for three or four years (Schlesinger et al. 1950; Shaw 1954).

Health organizations in 1951 were not as concerned as we are now about chronic exposure to trace poisons.

By the mid-1950s, a second scientific controversy occupied the nation, concerning harmful effects of radioactive fallout from nuclear weapons testing in the atmosphere. Stopping testing was a liberal cause. Since the protests against fluoridation and weapons testing occupied opposite ends of the political spectrum, few activists belonged to both movements, yet their risk messages were essentially the same. Both objected to involuntary chronic exposure of large populations to low doses of agents that were known to be very dangerous at higher doses. Both regarded distant and misguided leaders of government and industry as the responsible parties placing populations at risk. Both accused these parties of ignoring accumulating scientific evidence of chronic toxicity from low-level exposure. Both envisioned the poisons emanating from technology as insidiously contaminating the purity of nature. Both emphasized the process of bio-concentration, by which some trace poisons become increasingly concentrated as they are consumed by species higher up the food chain. Both saw in chemical pollution a symptom of the moral decay of society. Both worried particularly about cancer. The arguments against fluoridation and radiation are so similar as to be virtually interchangeable. These are, furthermore, exactly the elements that constitute the ideology of Rachel Carson's *Silent Spring,* which in 1962 would warn of DDT and other pesticides. Despite the commonality of messages, there was an intellectual disdain for antifluoridationists that never extended to opponents of atmospheric testing or of DDT, a difference perhaps reflecting the antifluoridationists' greater distance from the intellectual centers of the nation (Mazur 1998).

Risk Assessment

Much of the large literature evaluating fluoridation is frankly intended to either promote or discredit water treatment, so the consumer of research can locate reports of virtually any effect that is desired. A sensible overview of this work requires attention to reliable findings from methodologically strong studies, rather than anecdotal or anomalous claims, and the credibility of sources is important.

According to the Centers for Disease Control (CDC), early studies reported cavity reductions from fluoridation ranging from 50% to 70%, but studies during the 1980s showed reductions of only 8% to 37% among adolescents. This down trend has been attributed to the use of fluoride even in unfluoridated communities through bottled and processed food and beverages, and the use of fluoride toothpaste (CDC 2000).

Responding to a request from the U.S. Environmental Protection Agency (EPA) to determine whether its maximum contaminant level of 4 ppm fluoride in drinking water is appropriate, a subcommittee of the National Research Council, the principal operating agency of the National Academy of Sciences, in 1993 reviewed the health effects of ingested fluoride, providing the basis for the remainder of this section (NRC 1993).

The major sources of fluoride intake are water, soft drinks, and tooth paste. Children's exposure to fluoride has increased since the 1970s, even in non-fluoridated areas, due to its presence in so many products. Fluoride at recommended levels produces some dental fluorosis, usually as a barely discernible white spotting on the enamel but occasionally brown staining or pitting. The prevalence of mottling has increased, though there is disagreement whether in moderate form it is a health effect or a cosmetic problem.

The effect of fluoride on bone strength and hip fractures has been addressed in experimental studies on humans and animals, and in epidemiological comparisons of fracture rates in populations of elderly people that differed in their exposure to natural or added fluoride in drinking water. These yield inconsistent results, some showing a weak association between fluoride in drinking water and the risk of hip fracture. (There is little indication that fluoride strengthens bones.) In view of conflicting results and methodological weaknesses, the subcommittee found no basis for recommending EPA lower the current standard for fluoride but did recommend more research on fractures.

High exposures to fluoride are known to cause a variety of adverse health effects in experimental animals, but the subcommittee found no indication exposure below the EPA contaminant level of 4 ppm produced kidney disease, gastrointestinal or immune system problems, adverse reproductive effects, or genotoxicity.

More than 50 epidemiological studies have examined the relation between fluoride in drinking water and human cancer, most comparing geographic or temporal patterns of cancer rates with fluoride levels. This body of work had already been reviewed by several independent expert panels of epidemiologists, including the International Agency for Research on Cancer (IARC 1982), so the subcommittee elected to summarize prior findings, also considering eight recent studies, rather than undertaking another comprehensive review. The subcommittee reaffirmed earlier conclusions that this research provides no credible evidence for an association between fluoride in drinking water and the risk of cancer in humans; if a link exists it must be very weak. It also found available laboratory data insufficient to demonstrate a carcinogenic effect of fluoride in animals. Nonetheless, more and better-designed epidemiological research was recommended to more fully evaluate the relationship between fluoride exposure and cancer at various sites.

In denying any empirical link between fluoride and cancer, the subcommittee flatly contradicted studies by Yiamouyiannis and Burk (1977)—highly publicized by antifluoridationists—showing correlations among American cities between fluoride in water and cancer mortality. As noted by the IARC and other critics, those studies did not adjust adequately for age, race and sex in the groups that were being compared. For example, comparing fluoridated cities with older populations—hence high cancer mortality—with unfluoridated cities having younger populations—hence low cancer mortality—gives a spurious association of fluoridation with mortality, when in fact it is the age difference that explains the differing mortality rates.

Overall, the subcommittee found EPA's maximum contaminant level of 4 ppm fluoride in drinking water to be appropriate as an interim standard, pending new research results, while recognizing it would give a small percentage of the U.S. population moderate to severe dental fluorosis.

There have been infrequent reports of community outbreaks of acute fluoride poisoning due to overfluoridation of public water supplies. In one documented case, a faulty feed pump in a treatment plant of a small community allowed excessive fluoride into the tap water, one sample measuring 200 ppm. The day before the error was discovered, 14 people reported to the local hospital with acute nausea or vomiting (Penman et al. 1997).

Dynamics of Controversy

Now 50 years old, the fight over fluoridation continues and is by far the longest running technical controversy. Like classic theater, the polarized structure of its plot is invariant but the cast changes. As older actors die or retire, new players speak more or less the same lines, voicing arguments and rejoinders that mesh as if scripted (Martin 1991). Unlike a play, the actors hardly ever switch parts and there is no resolution.

Cogent evaluation of scientific evidence seemingly has little effect on partisan positions in a controversy as polarized as this one. A profluoridationist recently chastised opponents for linking water treatment "to a laundry list of aliments including ... even stained teeth," oblivious that stained teeth *is* a demonstrated consequence of fluoridation (Mahtesian 1997). On the other side, John Yiamouyiannis reiterated in a 1999 interview that fluoride "definitely" causes cancer (Chowka at <www.naturalhealthvillage.com/newsletter/990915/yiamouyiannis>).[1] Stalwarts fit any new evidence into prior conclusions. The opposition journal *Fluoride*, editorializing on new studies showing neurotoxicity in rats fed water with low concentrations of fluoride, noted "paradoxically" the same studies found toxicity significantly lower as fluoride concentration increased; the editorial nonetheless concluded adverse effects are "clear-cut" (Spittle and Burgstahler 1998).

Recalcitrants need not be poor scientists. Albert Einstein famously opposed quantum theory, denying "God plays dice with the universe" until the day he died, perhaps then learning the truth. Intransigence reminds us scientific evidence is never irrefutable, scientific claims cannot be established beyond all logical doubt.

What does change from year to year is the intensity of controversy, whether measured by the number of active opponents, the amount of news coverage about fluoridation, or the percentage of people who say on opinion polls they oppose fluoridation. All indicators rise during years of national concern about larger issues relevant to antifluoridationists, for this is when activists and journalists are most readily energized. Periods when Americans were especially fearful of communism and socialized medicine in the United States, reflected in the popularity of Senator Joseph McCarthy in the early 1950s and the presidential candidacy of Barry Goldwater in 1964, were times of peak antifluoridation activity. Another peak, around 1970, was tied to the incipient Environmental Movement and its concern with trace "poisons" such as DDT, mercury, and fluorides (Mazur 1981). Like a surfer catching a wave, opposition rises with these larger concerns, then diminishes as they wane. In recent decades, fluoridation activity, pro and con, has been relatively quiet at the national level, though it continues to inflame individual communities and there remains high likelihood, given a referendum, for treatment to be rejected.

Half the U.S. population has fluoride added to its drinking water, about the same portion as in the 1970s, while another few percent have naturally fluoridated water. Martin estimated in 1991 about 100 million people outside the United States drank water with added fluoride (Martin 1991, 193–217); today the number may be as high as 300 million, but in any case water fluoridation has not become widespread. Several countries in Europe and Latin America add fluoride to table salt, a voluntary medium that seemingly evokes less dissent.

Note

1. Yiamouyiannis believed stress is probably the major contributing cause of cancer and is co-author with Dr. Peter Duesberg of a book, *AIDS: The Good News Is That HIV Doesn't Cause It* (1995), denying that HIV causes AIDS.

REFERENCES

Adair, G. 1996. *Thomas Alva Edison: Inventing the Electric Age*. New York: Oxford University Press.

AEC (Atomic Energy Commission). 1972. "AEC Studies Report from Kansas Geological Survey of Possible Sites for Federal Waste Repository." AEC press release P–20. January 21.

Akin, W. 1977. *Technocracy and the American Dream*. Berkeley, CA: University of California Press.

Allwood, J. 1977. *The Great Exhibitions*. London: Studio Vista.

Altman, L. 1971. Football Injuries Are Linked to Synthetic Turf. *New York Times*, September 1: 39.

American Dental Association. 1965. Comments of the Opponents of Fluoridation. *Journal of the American Dental Association* 71: 1156.

Andervont, H. 1952. Testimony in *Hearings Before the House Select Committee to Investigate the Use of Chemicals in Foods and Cosmetics*, 1666–1667. 82nd Congress. Washington, DC.

Arkright, F. 1933. *The ABC of Technocracy*. New York: Harper & Row.

Armstrong, W., J. Bittner, and A. Treloar. 1954. Testimony in *Hearings on H.R. 2341 before the Committee on Interstate and Foreign Commerce: Fluoridation of Water*, 307–309. 83rd Congress. Washington, DC.

Beard, Charles. 1930. *Toward Civilization*. New York: Longmans, Green and Co.

Beck, U. 1992. *Risk Society*. Beverly Hills, CA: Sage.

Berton, P. 1992. *Niagara*. New York: Kodansha.

Boffey, P. 1970. Gofman and Tamplin: Harassment Charges Against AEC, Livermore. *Science* 169: 838.

———. 1975. *The Brain Bank of America*. New York: McGraw-Hill.

Bond, Victor. 1970. *Radiation Standards, Particularly as Related to Nuclear Power Plants*. Raleigh, NC: Council for the Advancement of Science Writing.

Bopp, B., R. Sonders, and J. Kesterson. 1986. Toxicological Aspects of Cyclamate and Cyclohexylamine. *CRC Critical Reviews of Toxicology* 16: 213.

Breyer, S. 1993. *Breaking the Vicious Circle*. Cambridge, MA: Harvard University Press.

Brown, Phil. 1991. The Popular Epidemiology Approach to Toxic Waste Contamination. In S. Couch and J. Kroll-Smith (eds.), *Communities at Risk*. New York: Peter Lang, 133–152.

Brusick, B., M. Cifone, R. Young, S. Benson. 1989. Assessment of the Genotoxicity of Calcium Cyclamate and Cyclohexylamine. *Environmental and Molecular Mutagenesis* 14: 188–199.

Buchanan, R. 2001. Email message to author.

Burford, Anne, and John Greenya. 1986. *Are You Tough Enough?* New York: McGraw-Hill.

Calorie Control Council and Abbott Laboratories. 1982. *Food Additive Petition for Cyclamate, 2A3672*. U.S. Food and Drug Commission.

Carson, R. 1962. *Silent Spring*. Boston: Houghton Mifflin.

Centers for Disease Control. 2000. Achievements in Public Health, 1900–1999: Fluoridation of Drinking Water to Prevent Dental Caries. *Journal of American Medical Association* 283: 1283–1286.

Crain, R., E. Katz, and D. Rosenthal. 1969. *The Politics of Community Conflict.* Indianapolis: Bobbs-Merrill.

Curley, A., R. Hawk, R. Kimbrough, G. Nathenson, and L. Finberg. 1971. Dermal Absorption of Hexachlorophane in Infants. *Lancet,* August 7: 296–297.

Dean, H. 1938. Endemic Fluorosis and Its Relation to Dental Caries. *Public Health Reports* 53: 1443–1452.

Delaney, J. 1957. Chemical Additives in Our Food Supply Can Cause Cancer. *Congressional Record.* Appendix. Feb. 21: A1351–A1354.

———. 1975. Fluoridation and Cancer. *Congressional Record,* 94th Congress, vol. 121: 23729–23733.

Dick, R. n.d. *Injury Surveillance System: 1997–98 Football.* Overland Park, KS: NCAA.

Driesen, D. 2001. Getting Our Priorities Straight: One Strand of the Regulatory Reform Debate. *Environmental Law Reporter* 31: 10003–10020.

Easterbrook, G. 1995. *A Moment on the Earth: The Coming Age of Environmental Optimism.* New York: Viking.

Efron, E. 1984. *The Apocalyptics.* New York: Simon and Schuster.

Elsner, J., Jr. 1967. *The Technocrats.* Syracuse, NY: Syracuse University Press.

Engelhardt, H., and A. Caplan 1987. *Scientific Controversies: Case Studies in the Resolution and Closure of Disputes in Science and Technology.* New York: Cambridge University Press.

EPA (Environmental Protection Agency). 1982. *Unfinished Business.* Washington, DC.

———. 2001. *Water Quality Criterion for the Protection of Human Health: Methylmercury.* Washington, DC: EPA.

Epstein, S. 1970. NTA. *Environment* 12: 3–11.

———. 1972. Toxicological and Environmental Implications of the Use of Nitrilotriacetic Acid as a Detergent Builder. *International Journal of Environmental Studies* 3: 13–21.

———. 1998. *The Politics of Cancer Revisited.* Fremont Center, NY: East Ridge Press.

Evans, R. 1966. The Effect of Skeletally Deposited Alpha Emitters in Man. *British Journal of Radiology* 39: 881–895.

Exner, F., G. Waldbott, and J. Rorty. 1957. *The American Fluoridation Experiment*. New York: Devin–Adair.

FDA (Food and Drug Administration). 1995. *FDA Backgrounder: Monosodium Glutamate*. Washington, DC: FDA.

FDA Science Board. 1997. *The Science Board Subcommittee on FDA Research Final Draft Report*. Washington, DC: FDA.

Flemming, B. 1991. *Arthur Flemming: Crusader at Large*. Washington, DC: Caring.

Forsyth, A. 1952. Testimony in *Hearings Before the House Select Committee to Investigate the Use of Chemicals in Foods and Cosmetics*, 1666–1667. 82nd Congress. Washington, DC.

Gillette, R. 1972. Radiation Standards: The Last Word or at Least a Definitive One. *Science* 178: 966.

Gladwell, M. 1989. U.S. Expected to Lift Ban on Cyclamate. *Washington Post*, May 16, A1.

Gofman, J., and A. Tamplin. 1971. *Poisoned Power*. Emmaus, PA: Rodale Press, 104–105.

Goldman, M. 1996. Cancer Risk of Low-Level Exposure. *Science* 271: 1821–1822.

Gots, R. 1993. *Toxic Risks: Science, Regulation, and Perception*. Boca Raton, FL: Lewis.

Graham, J., and J. Wiener. 1995. *Risk versus Risk*. Cambridge: Harvard University Press.

Greer, L. 2002. Email message to author.

Hamilton, R., and L. Hargens. 1993. The Politics of the Professors: Self Identification, 1969–1984. *Social Forces* 71: 603–627.

Hardin, Garrett. 1968. The Tragedy of the Commons. *Science* 162: 1243–1248.

Holcomb, R. 1970. Radiation Risk: A Scientific Problem? *Science* 167: 854.

Howe, H., P. Wingo, M. Thun, L. Ries, H. Rosenberg, E. Feigal, and B. Edwards. 2001. Annual Report to the Nation on the Status of Cancer

(1973–1998), Featuring Cancers With Recent Increasing Trends. *Journal of the National Cancer Institute* 93: 824–842.

Hughes, T. 1975. *Changing Attitudes toward American Technology*. New York: Harper and Row.

Hurley, P. 1998. Mode of Carcinogenic Action of Pesticides Inducing Thyroid Follicular Cell Tumors in Rodents. *Environmental Health Perspectives* 106: 437–444.

Huxley, A. 1932. *Brave New World*. New York: Harper and Row.

IARC (International Agency for Research on Cancer). 1987. Inorganic Fluorides. *IARC Monographs on the Evaluation of Carcinogenic Risk of Chemicals to Humans* 27: 235–303.

IPCC (Intergovernmental Panel on Climate Change). 2001. *Climate Change 200: Synthesis Report*. New York: Cambridge University Press.

Kokoski, C., S. Henry, C. Lin, and K. Ekelman. 1990. Methods Used in Safety Evaluation. In A. Branen, P. Davidson, and S. Salminen (eds.), *Food Additives*. New York: Marcel Dekker, 579–616.

Kopp, C. 1979. The Origins of the American Scientific Debate over Fallout Hazards. *Social Studies of Science* 9: 403–422.

Krimsky, S. 1991. *Biotechnics & Society*. New York: Praeger.

Krimsky, S., and D. Golding. 1992. *Social Theories of Risk*. Westport, CT: Praeger.

Ladd, E., and S.M. Lipset. 1976. *The Divided Academy*. New York: Norton.

Lawless, Edward. 1977. *Technology and Social Shock*. New Brunswick, NJ: Rutgers University Press.

Leake, C. 1966. Dimethyl Sulfoxide. *Science* 152: 1646–1649.

Levy, M., M. Skovron, and J. Agel. 1990. Living with Artificial Grass, Part 1. *American Journal of Sports Medicine* 18: 406–412.

Lichter, S., and S. Rothman. 1999. *Environmental Cancer: A Political Disease?* New Haven, CT: Yale University Press.

Little, A. 1924. The Fifth Estate. *Atlantic Monthly* 134: 771–781.

Lu, F. 1995. A Review of the Acceptable Daily Intakes of Pesticides Assessed by WHO. *Regulatory Toxicology and Pharmacology* 21: 352–364.

Mahtesian, C. 1997. Tooth Squads. *Governing* (June): 40.

Marco, G., R. Hallingworth, and W. Durham (eds.). 1987. *Silent Spring Revisited*. Washington, DC: American Chemical Society.

Marcus, A. 1994. *Cancer from Beef: DES, Federal Food Regulation, and Consumer Confidence*. Baltimore: Johns Hopkins University Press.

Martin, B. 1991. *Scientific Knowledge in Controversy*. Albany, NY: State University of New York Press.

Mausner, B., and J. Mausner. 1955. A Study of the Anti-Scientific Attitude. *Scientific American* 192: 35–39.

Mazur, A. 1981. *The Dynamics of Technical Controversy*. Washington, DC: Communications Press.

———. 1984. The Journalists and Technology: Reporting about Love Canal and Three Mile Island. *Minerva* 22: 45–66.

———. 1985. Bias in Risk-Benefit Analysis. *Technology in Society* 7: 25–30.

———. 1986. U.S. Trends in Feminine Beauty and Overadaptation. *Journal of Sex Research* 22: 281.

———. 1987. Putting Radon on the Public's Risk Agenda. *Science, Technology, and Human Values* 12: 86–93.

———. 1990. Nuclear Power, Chemical Hazards, and the Quantity of Reporting. *Minerva* 28: 294–323.

———. 1996. Why Do We Worry About Trace Poisons? *Risk: Health, Safety & Environment*: 35.

———. 1998. *A Hazardous Inquiry: The Rashomon Effect at Love Canal*. Cambridge, MA: Harvard University Press.

———. 2000. Looking Back: Unneeded X-Rays. *Health, Safety & Environment* 11 (winter): 1–8.

———. 2001. Looking Back at Fluoridation. *Risk: Health, Safety & Environment* 12 (spring): 59–65.

Mazur, A., and J. Bretsch. 1999. Looking Back: Synthetic Turf and Football Injuries. *Risk: Health, Safety & Environment* 10 (winter): 1–6.

Mazur, A., and K. Jacobson. 1999. Looking Back: Cyclamate. *Risk: Health, Safety & Environment* 10 (spring): 95–100.

Mazur, A., and J. Lee. 1993. Sounding the Global Alarm: Environmental Issues in the U.S. National News. *Social Studies of Science* 23: 681–720.

Mazur, A., S. Rothman, and S. Lichter. 2001. Biases about Man-Made Cancer among Researchers. *Social Studies of Science* 31(5): 771–778.

McCarry, Charles. 1972. *Citizen Nader*. New York: New American Library.

McClure, F. 1970. *Water Fluoridation: The Search and the Victory*. Bethesda, MD: U.S. Department of Health, Education, and Welfare.

McFarland, A. 1976. *Public Interest Lobbies*. Washington, DC: American Enterprise Institute.

McGrady, P. , Sr. 1973. *The Persecuted Drug: The Story of DMSO*. Garden City, NY: Doubleday.

McNeil, D. 1985. America's Longest War: The Fight over Fluoridation, 1950–.*Wilson Quarterly* 9(3): 140–153.

McPhee, John. 1971. *Encounters with the Archdruid*. New York: Farrar, Straus and Giroux.

Meister, Kathleen. 1993. *Low-Calorie Sweeteners*. New York: American Council on Science and Health.

Millikan, R. 1930. Science Lights the Torch. In C. Beard (ed.), *Toward Civilization*. New York: Longmans, Green and Co., 38–46.

Mintz, M. 1969. Rise and Fall of Cyclamates. *Washington Post* (Oct. 26): A1.

Mumford, Lewis. 1928. The Arts. In C. Beard (ed.), *Whither Mankind?* New York: Longmans, Green and Co., 287–312.

———. 1934. *Technics and Civilization*. New York: Harcourt, Brace and Co.

Nader, R. 1965. *Unsafe at Any Speed*. New York: Grossman.

Nathanson, N., and A.D. Langmuir. 1963. The Cutter Incident: Poliomyelitis Following Formaldehyde-Inactivated Poliovirus Vaccination in the United States During the Spring of 1955. II. Relationship of Poliomyelitis to Cutter Vaccine. *American Journal of Epidemiology* 142(2):109–140.

National Public Radio. Morning Edition. 2001. April 27.

NAS (National Academy of Sciences). 1956. *Biological Effects of Atomic Radiation*. Washington, DC.

———. 1972. *The Effects on Populations of Exposure to Low Levels of Ionizing Radiation*. Washington, DC.

———. 1973. Dimethyl Sulfoxide as a Therapeutic Agent. Washington, DC.

———. 1985. *Evaluation of Cyclamate for Carcinogenicity*. Washington, DC.

———. 1990. *Health Effects of Exposure to Low Levels of Ionizing Radiation: BEIR V*. Washington, DC.

National Academy of Sciences–National Research Council. 1973. *Dimethyl Sulfoxide as a Therapeutic Agent*. Washington, DC: NAS.

National Committee on Radiation Protection. 1999. *The Application of ALARA for Occupational Exposures*. Washington, DC.

National Research Council. 1993. *Health Effects of Ingested Fluoride*. Washington DC: National Academy of Sciences.

———. 1994. *Judgment in Risk Assessment*. Washington DC: National Academy of Sciences.

Nelson, B. 1970. Colorado Environmentalists: Scientists Battle AEC and Army. *Science* 168: 1324–1328.

———. 1971. Mobile TB X-Ray Units: An Obsolete Technology Lingers. *Science* 174: 1114.

New York Times. 1949. Doctors to Study DDT as Food Poison. April 3, 53.

———. 1952. Fight over Fluoride. March 10.

———. 1955. One Firm's Vaccine Barred; 6 Polio Cases Are Studied. April 28, 1.

———. 1956a. Hormone and Cancer. January 29, E9.

———. 1956b. Scientists Term Radiation a Peril to Future of Man. June 13, 1.

———. 1957. U.S. Denies Charge on Food Content. February 23, 10.

———. 1959. Some of Cranberry Crop Tainted By a Weed-Killer, U.S. Warns. November 10, 1.

———. 1962. Whole Fish Flour Is Barred for U.S. January 25, 28.

———. 1965. DMSO—Promise and Danger. April 3, 28.

———. 1967a. General Electric Will Modify 90,000 Large Color Television Sets against Possible X-Radiation Leaks. May 19, 17.

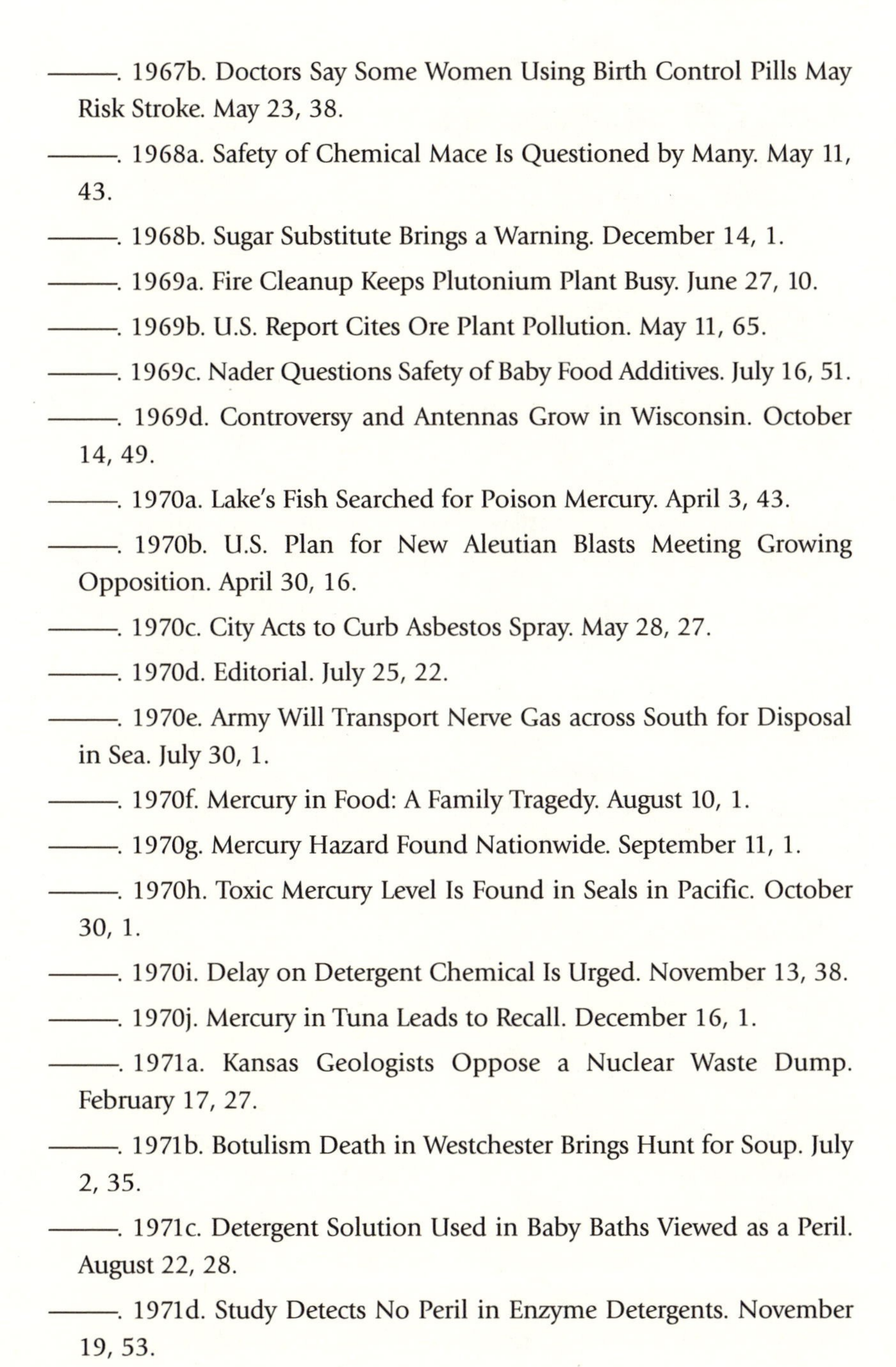

———. 1967b. Doctors Say Some Women Using Birth Control Pills May Risk Stroke. May 23, 38.

———. 1968a. Safety of Chemical Mace Is Questioned by Many. May 11, 43.

———. 1968b. Sugar Substitute Brings a Warning. December 14, 1.

———. 1969a. Fire Cleanup Keeps Plutonium Plant Busy. June 27, 10.

———. 1969b. U.S. Report Cites Ore Plant Pollution. May 11, 65.

———. 1969c. Nader Questions Safety of Baby Food Additives. July 16, 51.

———. 1969d. Controversy and Antennas Grow in Wisconsin. October 14, 49.

———. 1970a. Lake's Fish Searched for Poison Mercury. April 3, 43.

———. 1970b. U.S. Plan for New Aleutian Blasts Meeting Growing Opposition. April 30, 16.

———. 1970c. City Acts to Curb Asbestos Spray. May 28, 27.

———. 1970d. Editorial. July 25, 22.

———. 1970e. Army Will Transport Nerve Gas across South for Disposal in Sea. July 30, 1.

———. 1970f. Mercury in Food: A Family Tragedy. August 10, 1.

———. 1970g. Mercury Hazard Found Nationwide. September 11, 1.

———. 1970h. Toxic Mercury Level Is Found in Seals in Pacific. October 30, 1.

———. 1970i. Delay on Detergent Chemical Is Urged. November 13, 38.

———. 1970j. Mercury in Tuna Leads to Recall. December 16, 1.

———. 1971a. Kansas Geologists Oppose a Nuclear Waste Dump. February 17, 27.

———. 1971b. Botulism Death in Westchester Brings Hunt for Soup. July 2, 35.

———. 1971c. Detergent Solution Used in Baby Baths Viewed as a Peril. August 22, 28.

———. 1971d. Study Detects No Peril in Enzyme Detergents. November 19, 53.

———. 2001. FDA Warns Women Not to Eat Some Fish. January 14, 22.

Nicholas, J., P. Rosenthal, and G. Gleim. 1988. A Historical Perspective of Injuries in Professional Football. *Journal of the American Medical Association* 260: 939–944.

Nigg, B., and B. Segesser. 1988. The Influence of Playing Surfaces on the Load on the Locomotor System and on Football and Tennis Injuries. *Sports Medicine* 5: 375–385.

Novick, S. 1969. *The Careless Atom*. Boston, MA: Dell.

NRC (Nuclear Regulatory Commission). 1997. *Possible Health Effects of Exposure to Residential Electric and Magnetic Fields.* Washington, DC: NRC.

———. 1999. *Research on Power-Frequency Fields Completed Under the Energy Policy Act of 1992*. Washington, DC: NRC.

Olney, J. 1969. Brain Lesions, Obesity, and Other Disturbances in Mice Treated with Monosodium Glutamate. *Science* 164: 719–721.

Pauling, L. 1958. *No More War!* New York: Dodd, Mead.

Paynter, O., G. Burin, R. Jaeger, and C. Gregorio. 1988. Goitrogens and Thyroid Follicular Cell Neoplasia: Evidence for a Threshold Process. *Regulatory Toxicology and Pharmacology* 8: 102–119 [review].

Penman, A., B. Brackin, and R. Embrey, 1997. Outbreak of Acute Fluoride Poisoning Caused by a Fluoride Overfeed, Mississippi, 1993. *Public Health Reports* 112: 403–409.

Perlmutter, E. 1962. Doctor's Action Bars Birth Defects. *New York Times,* July 16, 24.

Petroski, H. 1997. *Remaking the World*. New York: Alfred Knopf.

Powell, J., and M. Schootman. 1992. A Multivariate Risk Analysis of Selected Playing Surfaces in the National Football League: 1980 to 1989. *American Journal of Sports Medicine* 20: 686–694.

Powell, M. 1999. *Science at EPA: Information in the Regulatory Process.* Washington, DC: Resources for the Future.

President's Commission on the Accident at Three Mile Island. 1979. *Report of the Public's Right to Information Task Force*. Washington, DC: U.S. Government Printing Office.

Rapoport, Roger. 1968. MACE in the Face. *New Republic* 158: 14.

Reuters. 1999. EU Rejects Swedish Restrictions on Food Additives. January 8.

Rodeo, S.A., S. O'Brien, R.F. Warren, R. Barnes, T.L. Wickiewicz, and M.F. Dillingham. 1990. Turf-Toe: An Analysis of Metatarsophalangeal Joint Sprains in Professional Football Players. *American Journal of Sports Medicine* 18 (3): 280–285.

Roueche, B. 1953. *Eleven Blue Men*. Boston: Little, Brown.

Russell, B. 1928. Science. In C. Beard (ed.), *Whither Mankind?* New York: Longmans, Green and Co., 63–83.

Sandman, P., and M. Paden. 1979. At Three Mile Island. *Columbia Journalism Review* (July/August): 43–58.

Schiefelbein, S. 1979. The Invisible Threat. *Saturday Review* (Sept. 15): 16–20.

Schlesinger, E., D. Overton, and H. Chase, 1950. Newburgh–Kingston Caries–Fluorine Study. *American Journal of Public Health* 40: 725–727.

Science Board Subcommittee on FDA Research. 1997. *Recommendations to the Science Board of the Food and Drug Administration*. Washington, DC: FDA.

Shaw, J. 1954. *Fluoridation as a Public Health Measure*. Washington, DC: American Association for the Advancement of Science.

Sinclair, Upton. 1906. *The Jungle*. New York: Doubleday, Page.

Singer, E., and P. Endreny. 1994. *Reporting on Risk*. New York: Russell Sage Foundation.

Skovron, M., M. Levy, and J. Agel. 1990. Living with Artificial Grass: Part 2. *American Journal of Sports Medicine* 18: 510–513.

Speirs, A. 1962. Thalidomide and Congenital Abnormalities. *Lancet* (Feb. 10): 1962, 303–305.

Spittle, B., and A. Burgstahler, 1998. Death Knell for Fluoridation? *Fluoride* 31: 59–60.

Stallings, R. 1994. Hindsight, Organizational Routines, and Media Risk Coverage. *Risk: Health, Safety & Environment* 5: 271–280.

Stegner, Wallace. 1983. The Best Idea Ever. *Wilderness* 47 (spring).

Stern, P., and J. Fineberg (eds.). 1996. *Understanding Risk: Informing Decisions in a Democratic Society*. Washington, DC: National Research Council, NAS.

Tamplin, Arthur, and John Gofman. 1970. *Population Control through Nuclear Pollution*. Chicago: Nelson-Hall.

Taylor, L. 1958. History of the International Commission on Radiological Protection. *Health Physics* 1: 97.

Time. 1949. Little Feet, Be Careful. September 9, 67.

———. 1962. Sleeping Pill Nightmare. February 23, 86.

———. 1970. Enzymes in Hot Water. February 16, 86.

Toffler, Alvin. 1970. *Future Shock*. New York: Random House.

Turner, J. 1970. *The Chemical Feast*. Washington, DC: Center for Study of Responsive Law.

Veblen, Thorstein. 1921. *The Engineers and the Price System*. New York: B.W. Huebach.

Verrett, M., W.F. Scott, E.F. Reynaldo, E.K. Alterman, C.A. Thomas. 1980. Toxicity and Teratogenicity of Food Additive Chemicals in the Developing Chicken Embryo. *Toxicology and Applied Pharmacology* 56(2): 265–273.

Wade, N. 1971. Hexachlorophene: FDA Temporizes on Brain-Damaging Chemical. *Science* 174: 805–807.

Wallace, L. 1930. Engineering in Government. In C. Beard (ed.), *Toward Civilization*. New York: Longmans, Green and Co., 176–195.

Wall Street Journal. 1959. Chemical That Causes Cancer in Animals Is Found in Some West Coast Cranberries. November 10, 3.

Warren, C. 2000. *Brush with Death*. Baltimore, MD: Johns Hopkins University Press.

Webster, P. 1972. *The Mighty Sierra*. New York: Weathervane Books.

Weinberg, Alvin. 1972. Science and Trans-Science. *Minerva* 10: 209–222.

Whelan, E. 1993. *Toxic Terror*. Buffalo: Prometheus Books.

Williams, Gary (ed.). 1988. *Sweeteners: Health Effects*. Princeton, NJ: Princeton Scientific Publishing Co.

Wilson, J. 1963. *Margin of Safety*. Garden City, NY: Doubleday.

Yiamouyiannis, J., and D. Burk. 1977. Fluoridation and Cancer: Age-Dependence of Cancer Mortality Related to Artificial Fluoridation. *Fluoride* 10: 102–125.

INDEX

About the Author

Allan Mazur, a sociologist and engineer, is a professor of public affairs in the Maxwell School of Syracuse University. His research focuses on social aspects of science, technology, and the environment, and on biosociology. He is especially interested in disputes between experts about risks to health, and how they are reported in the mass media. Mazur is a fellow of the American Association for the Advancement of Science and was recently Gilbert White Fellow at Resources for the Future. He is author or coauthor of over 150 articles and five books. His most recent book, *A Hazardous Inquiry,* looks back at the controversy that erupted over the infamous toxic waste site at Love Canal.